WATER DEVELOPMENT AND MANAGEMENT IN UTTAR PRADESH, INDIA

PAST, PRESENT, AND FUTURE PERSPECTIVE

Pratik Ranjan Chaurasia & Ravindra Swaroop Sinha

Contents

Summary

Uttar Pradesh is one of the states in India where tremendous water development has taken place after independence for extension of irrigation facilities and fulfilling other requirements. At present state boasts to have about 87 % irrigated area while the national average is only about 49 %. Before five year plan period, irrigation potential created was 5.4 million hectares only, which is now more than 36 million hectares, though there is a huge gap between potential created and potential utilized. The net irrigated area also increased from about 3.2 million hectares to about 14.4 million hectares after independence. Thus, the state is now categorized as one of the highly irrigated states of India. After independence, drinking water, and industrial facilities also increased manifold in the state to fetch the growing demand of its population. With an increase in irrigation facilities crop productivity and food production also increased manifold in the state and now the state is one of the major contributors to the food basket of the Nation.

In its endeavor for water development to meet the growing demand of the growing population, the state somewhere ignored the management aspects of water development and relied too much on groundwater as the state is comparatively rich in groundwater resource which is easily available and its abstraction is very easy and cheap in the state. The invention of low-cost pump sets in the 1980s started the tube well revolution in the country and Uttar Pradesh became the center of the tube well revolution. Though the state has a large canal network, irrigation is mainly dependent on groundwater. There are about 4 million tube wells and wells in the state providing 81 % irrigation supplies. Apart from this, groundwater is fulfilling about 80 % of drinking water requirements and 90 % of industrial demand. This has led to heavy exploitation of this precious natural resource. Due to excessive abstraction of groundwater, out of 820 developmental blocks, the water level is declining in 572 blocks, which indicates groundwater mining or abstraction of non-renewable groundwater in a majority of blocks.

In most of the 653 major and medium townships in the state, groundwater levels are also declining at alarming rates as easy resource availability and access has encouraged mushrooming tube well construction activity. Rainfall pattern is also changing in the state due to climate change complications and rainfall is not a reliable source of water supplies now. In the last three decades, the annual average rainfall is continuously declining, triggering water crises in the state. River basins are water-short now indicating future water crises and call for demand-side interventions. Canal commands continue to face the problem of waterlogging while tail ends face the non-availability of water.

Though, of late, the state started to pay attention to the management aspect of water development through bringing "State Water Policy, 1999", "Overall Policy for Groundwater Management, Rainwater Harvesting and Groundwater Recharge, 2013", enacting "Participatory Irrigation Management Act 2009", Water Management and Regulatory Commission Act 2014", "Groundwater (Management and Regulation) Act, 2019", modifying building by-laws for rainwater harvesting, re-organizing departments dealing with water and constituting unified Water Power Ministry, etc., State has to go a long way to manage its water resources effectively.

The book discusses in detail the surface and groundwater availability and their development in the state, irrigation facilities, and effect on productivity due to extension of irrigation facilities, present and future availability of water, future needs, the status of water management interventions, present challenges such as declining rainfall, underutilization of irrigation potential, water scarcity, water quality, water laws, and governance, water efficiencies, etc. and future interventions to manage water crises and fulfill future demands, etc. It also discusses water crises in urban areas and required policy initiatives/ measures.

Keywords: Water resources, Surface water, Groundwater, Non-renewable groundwater, Major & Medium irrigation schemes, Minor irrigation schemes, Dynamic groundwater resource, Groundwater level decline, Urban water supply, Over-exploited, Critical and Semi-critical areas, River, River basin, Aquifer system/ Groundwater basin, Groundwater, and Surface water monitoring, Digital or Automatic Groundwater Level recorder, Real-time monitoring, Telemetry, Barrage, Dam, Irrigation sources, Irrigated area, Micro-irrigation, Irrigation potential and utilization, Minor irrigation census, Waterlogging, Command area, Crop production, Crop productivity, Water resources management, Water governance, Water policy, Water acts, Organizational change, World Bank schemes, Rainfall, Forest cover, Conjunctive use, Water efficiency, Integrated planning, Wastewater, Productivity of water, Water quality, Traditional water bodies/ponds, Internet of things, Automatization, Groundwater recharge/restoration.

Tables and Figures

Tables

Figures

Foreword

The book is focused on how, over the years water was developed and managed in the state, challenges the water sector faces at present and future interventions to counter these challenges. Uttar Pradesh (UP) is the highest populated state of the country. Though its geographical area is only about 8 % of the country's geographical area, it supports about 17 % population of the country, merely because it is rich in water resources; surface, and underground. Water is necessary for life as well as crucial for societal growth. Without it, no developmental activity can take place. A major part of the state lies in the fertile Ganga and Yamuna plains where the high depth of alluvium or sedimentary deposits provides a large space for groundwater storage. Some of the major rivers of the country flow through the state.

With the advancement of technology, the state has witnessed very high activity for the development of water in the last four decades bringing the state into the highly irrigated category which calumniated in to increase in food production. At present, the state is one of the major contributors to the food basket of the country. Though with a growing population, its water demand is also growing and its major river basins are becoming water short. In the majority of blocks, groundwater levels are declining indicating groundwater mining or abstraction of non-renewable groundwater. In the last three decades, the rainfall patterns show a declining trend having erratic behavior. Thus despite being rich in water resources, the state, now is in danger of future water crises and faces new water management challenges. The state will be able to fulfill future water demands only if, the ways water is managed and governed in the state at present are changed drastically.

The book is divided into 10 chapters. Chapter 1 deals with the water resource of the state and their level of exploitation that includes availability and utilization of surface and groundwater, stressed areas, etc. Chapter 2 discusses the river system of the state, major rivers, basins, and dams/barrages. Chapter 3 is focused on the groundwater aquifer system of the state and discusses groundwater occurrence, aquifers in alluvium, and rocky areas. Chapter 4 is dedicated to the surface water and groundwater monitoring infrastructure in the state. Chapter 5 is regarding the development of surface water and groundwater sources over the years and discusses the development of irrigation sources like canals, government tube-wells, private tube-wells, dug wells, etc. Chapter 6 describes the growth of the irrigated area and crop production including source-wise irrigated area, irrigated area through surface and groundwater resources, coverage of modern irrigation systems, irrigation potential and utilization, the extent of waterlogged areas, etc. Chapter 7 is about domestic water supplies. Chapter 8 deals with the efforts made by the state for water management and governance and discusses various policy initiatives, acts, executive orders, and schemes. Chapter 9 analyses various challenges the state is facing in the water sector at present like diminishing rainfall, inadequate forest cover, declining groundwater level, shallow groundwater level, water pollution, water efficiencies, etc. To face these challenges, chapter 10 suggests strategies and future interventions. Lastly, conclusions are narrated.

Pieces of information regarding the water sector are not usually on one platform. An effort has been made to consolidate various data and information related to the water sector development and present these in an integrated manner. I hope that the book will be useful to the policymakers, planners, researchers, and field-level functionaries engaged in water development and management.

Pratik Ranjan Chaurasia

6th January 2021

1. WATER RESOURCES OF THE STATE

1.1 Surface Water

The state is endowed with bountiful water resources. The assessment by the water resource organization, Government of India indicates that 75% dependability flow of five major rivers of the state is 99.06 million acre-feet (MAF) or 122.188 billion cubic meters (BCM) as shown in Table 1.1[1]. According to this assessment, the quantum of water available for exploitation in the state will be 99.06 MAF which is less than the annual flow of 145.01 MAF, including the share of other states. About 32 MAF water is already being used through various schemes in operation. Thus total surface water resource which can be developed by the state (including the state of Uttarakhand) is 131.06 (99.06+32) MAF. According to State Water Policy (SWP), 1999 also, the total surface water resource of UP including the state of Uttarakhand is 131.0 MAF of which 20 MAF is reserved for drinking water, industrial and pollution control, and 35 MAF is the quantity which cannot be utilized at present [2]. This leaves a balance of 76 MAF for irrigation.

Table 1.1.Water availability and annual flow

S.no.	Name of river	Discharge observation site	Annual flow (MAF)	Share of catchments of Uttar Pradesh (MAF)
1	2	3	4	5
1.	Ganga	Varansi	54.50	34.80
2.	Gandak	Balmikinagar	56.77	8.45
3.	Ghaghra	Turtipar	50.61	50.61
4.	Sone	Chopan	9.25	1.23
5.	Gomti	Naighat	3.97	3.97
	Total		145.10	99.06

Note: MAF= million acre feet, 01 MAF = 1.23348 BCM (billion cubic meters)

Tentative break up of ultimate irrigation potential likely to be created in UP and Uttarakhand through surface water is shown in Table 1.2 [1].

Table 1.2.Share of Surface Water Resources between Uttar Pradesh and Uttarakhand

Items	Ultimate Potential (million hectares)			Water Required (MAF)		
	Share of Uttar Pradesh	Share of Uttara khand State	Total	Share of Uttar Pradesh	Share of Uttara khand	Total
1	2	3	4	5	6	7
Major and Medium Irrigation	12.154	0.346	12.50	67.40	1.90	69.30
Minor Surface Irrigation	0.682	0.518	1.20	3.80	2.90	6.70
Total	12.836	0.864	13.70	71.20	4.80	76.0

Thus, according to Table 1.2, 71.20 MAF of water is left for UP through which ultimate irrigation potential of 12.836 million hectares; 12.154 million hectares from major and medium irrigation, and 0.682 million hectares from minor surface irrigation schemes can be created. The State Water Policy, 1999 also says that after completion of ongoing projects, only 18 MAF of total available water in the state or only 6.33 % will be left for future schemes. This means that almost all the available surface water had already been tapped through various schemes. The quantity of surface water which cannot be tapped at present i.e. 35 MAF is quite high, about 26.7 % of total available water.

1.2 Groundwater

The state of UP is very rich in groundwater resources and has a large groundwater reservoir. According to the latest assessment total dynamic groundwater resource of the state, as of 31.03.2017 is 69.92 BCM of which 65.32 BCM is extractable. Out of this 45.84 BCM is being abstracted including 4.9 BCM for drinking, domestic and industrial needs, leaving 20.36 BCM for future development, though, there seems some mathematical mistake as this figure comes out to be 19.48 BCM instead of 20.36 BCM. The stage of development or percent abstraction of annual recharge is 70.18 % [3].

Analysis of dynamic groundwater resources data, gathered from the last 10 periodic assessments carried out from the year 1975 to the year 2017, show significant changes in groundwater recharge and draft (extraction) components. A marked increase in groundwater extraction from the year 1995 is observed which is mainly attributed to the extensive groundwater development program in the state for meeting the growing demand for irrigation water in the agriculture sector. Data also show that groundwater extraction for domestic and industrial use increased too, after the year 2000. However, in extraction for industrial and domestic use, industrial use has not been estimated. According to guidelines, it should have been estimated.

Assessment of dynamic groundwater resource over the years is shown in Table 1.3[4] and figure 1, which shows that in the 2017 assessment, groundwater recharge as well as groundwater abstractions, have reduced by 8.7 % and 13.2 % respectively as compared to the 2013 assessment. Though one reason for the reduction in annual recharge may be due to the reduction in rainfall, it indirectly indicates that large-scale rainwater harvesting activities implemented in the state did not give the desired results. Similarly, reduction in groundwater abstraction is also against the general perception that uncontrolled abstraction of groundwater is rising, and groundwater levels are declining. According to the groundwater assessment report 2017 [3], out of 820 developmental blocks, the groundwater level is declining in 572 blocks indicating groundwater mining or abstraction of fossil groundwater or non-renewable groundwater which is not the active part of the current hydrological cycle and cannot be replenished. Importantly, when groundwater abstraction exceeds the recharge rates over extensive areas for prolonged periods, persistent groundwater depletion occurs leading to falling groundwater levels.

Table 1.3 Dynamic Groundwater Resource Assessment over the years

Billion Cubic Meters (BCM)

S.no.	Assessment Year	Groundwater recharge	Groundwater draft for all usages	Groundwater draft for industrial and domestic use
1	2	3	4	5
1.	1975	70.5	26.3	0.4
2.	1986	76.9	26.2	0.4
3.	1990	64.4	26.4	0.4
4.	1995	71.6	26.9	0.4
5.	2000	80.8	42.2	0.7
6.	2004	70.1	48.8	3.4
7.	2009	68.6	49.6	3.5
8.	2011	71.6	52.4	3.5
9.	2013	71.5	52.8	4.4
10.	2017	65.3	45.8	4.9

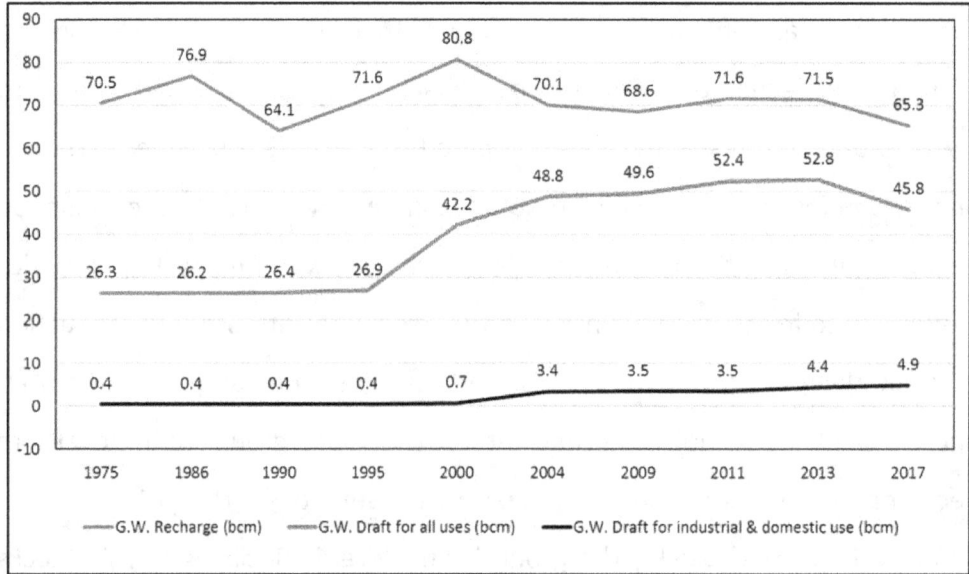

Fig. 1 Dynamic Groundwater Resource assessment over the years

Out of assessed 820 blocks, 280 blocks, or about 34 % are categorized as overexploited, critical, and semi-critical (OCS) categories. In the 2013 assessment, this number was 217 or there is an increase of about 29 %. The changing pattern of Over-Exploited (O E), Critical (C), and Semi-Critical (SC) areas or OCS blocks/areas are presented in Table 1.4 [3,5].

Table 1.4 Relative scenario of different categories of the stressed areas in Uttar Pradesh

Category	Blocks/Cities (Nos.)						
	Year 2000	Year 2004	Year 2009	Year 2011	Year 2013	The year 2017	
						Blocks	Cities
Over-exploited	2	37	76	111	113	82	9
Critical	20	13	32	68	59	47	1
Semi-critical	53	88	107	82	45	151	-
Sub-total	75	158	215	261	217	280	10
Safe	745	682	605	559	603	540	NA
Total	820	820	820	820	820	820	10

Note: NA-Not Available

Table 1.4 shows that the numbers of over-exploited, critical, and semi-critical areas are variably changing, influenced by the pattern of groundwater recharge and extraction, which is mostly non-uniform. Assessment of 2017 shows a remarkable jump in SC blocks and a decrease in OE blocks changing the trend. Overall, the numbers of OCS blocks are continuously on the rise, and now 34 % of blocks are stressed. Relative Increase in OCS areas is shown in Table 1.5[4].

Table 1.5 Percentage increase in stressed areas

Area	Over exploited, Critical, and Semi critical areas (nos.)		% increase
	The year 2013	The year 2017	
India	1968	2471	25.55
Uttar Pradesh	217	290	33.60

Table 1.5 suggests that the rate of increase in OCS areas is comparatively high in UP.

For the first time, Ground Water Resources Estimation Committee-2015 (GEC-2015) methodology [6] provides to assess groundwater resources in urban areas; however, this arrangement is made only on ad hoc norms. Ten cities having a population of more than 1 million were selected for the 2017 assessment. These are Lucknow, Kanpur Nagar, Aligarh, Agra, Ghaziabad, Meerut, Moradabad, Bareilly, Prayagraj, and Varanasi. GEC-2015 methodology states that since urban areas are concrete jungle, 30 % of the rainfall infiltration factor is considered for estimation of recharge component as an ad hoc arrangement till field studies are done. For recharge from other sources, seepages from pipeline leakages, sewage, flash floods are recommended for estimation purposes. In the absence of any census of groundwater extraction structures in cities, the available extraction data is not accurate. Therefore, the methodology has recommended taking the difference of the actual demand and the supply by surface water sources as the withdrawal from groundwater resources. In UP, surface water sources are not available in all the urban bodies, and the majority of such areas are dependent on groundwater, where extraction from water supply tube wells should be used for assessment. But, this aspect is probably ignored in the present resource estimation, which needs review and reassessment.

The city-wise resource estimation, as per GEC-2015, is given in Table 1.6 [3]. Table 1.6 suggests that the categorization of the mentioned cities is done based on the stage of development of groundwater resources without taking into account the declining or rising trend in groundwater level.

Further, the quantum of annual extraction is reported highest from Lucknow as 9656 ham or 264 million liters per day, which is on the lower side if the Lucknow Jal Sansthan's data of 390 million liters per day on tube well supply is compared. Prayagraj city is categorized as over-exploited, but the groundwater level is not declining in both the pre and post-monsoon periods. Similarly, in over-exploited Varanasi city, post-monsoon groundwater level shows no decline, while pre-monsoon groundwater level decline is of only 15 cm/year, which is less than the threshold of the yearly drop of 20 cm. Thus, the GEC-2015 methodology needs a review, whether areas can be categorized without taking into account the groundwater level trend. This becomes more essential because, at present, block-wise or social boundary-wise, not aquifer boundary-wise data are available for estimation of the groundwater resource. In contrast, aquifers do not follow social boundaries.

Table 1.6 Urban Groundwater Resource estimation 2017 (cities with population>1 million)

City	Area (km^2)	Extractable Ground Water Resource (ham)	Ground water Extraction (ham)	Stage of Extraction (%)	Decline (cm/year)		Category
					Pre mon.	Post mon.	
1	2	3	4	5	6	7	8
Agra	141.97	1402	1309	93	43	13	C
Aligarh	68.5	1040	3572	343	32	39	OE
Bareilly	134	2197	5111	232	15	52	OE
Ghaziabad	210	3461	9115	263	97	110	OE
Kanpur Nagar	278	4212	4311	102	51	47	OE
Lucknow	340	5451	9656	177	44	54	OE
Meerut	141	2136	5262	246	27	47	OE
Moradabad	91	1759	5414	308	27	34	OE
Prayagraj	82	2562	3810	149	Rising	Rising	OE
Varanasi	68	2445	4913	201	15	Rising	OE

Note OE- Over-exploited; C- Critical)

There is no authentic assessment of static groundwater resource or fossil groundwater resource or non-renewable groundwater resource in the state. This resource is not an active part of the current hydrological cycle but is a reserve and acts as an additional source of water in the dry season or dry years. Though GEC-15 methodology has provision for assessing static groundwater resources also, no attempt has yet been made to assess this resource. However there is a consensus that UP also has a sizeable static groundwater resource, but this resource should be used very cautiously and under emergencies only.

2. RIVER SYSTEM OF THE STATE

Most of the major rivers of the state either originate from the Himalayas or Vindhyans and the majority of them are perineal. Some rivers also originate from lakes and springs. The origin of major rivers is shown in Table 2.1 [7].

Table 2.1 Origin of Major rivers

S. no.	Name of the rivers	Origin
1	2	3
1.	Ganga, Yamuna, Ramganga, Sharada, Ghagra, Gandak, Rapti, Banganga, Hindon, Sarju, Kosi, Gaura, rohani etc.	Himalayas
2.	Betwa, Dhasan, Sindh, Son, Rehand, Ken, Tons, Kanhar, Belan, Karmnasa, Khari, Shahjad, Sajnam, Jamni, Pahuj, Rohani, Lakheri, Birma,Urmil, Chandrawal, Arjun, Baghain, Chambal etc.	Vindhyans
3.	Gomti, Sai, Sengar, Arind, Jhirna, Non, Pandu, Ishan, Behta, Kalyani, Bakulahi, Varuna etc.	Springs and lakes

2.1 Rivers and River Basins

River Ganga and its tributaries drain the entire state. Prominent rivers of the state and the detailed basins are shown in Table 2.2[7] and figure 2.

The River Ganga originates from the great Himalayan range and runs over a distance of 1400 km in the state. Ramganga, Gomti, Ghaghra, Yamuna, Betwa, Tons, Son, are some of its major tributaries. The main left-bank tributaries of Ganga are Ramganga, Gomti, while the main right-bank tributary is the Yamuna River. It is the most important and sacred river.

The Ramganga, left-bank tributary of Ganga, originates from the Nag Tiba Range of the Himalayan Region, and after flowing easterly over a distance of 596 km, joins River Ganga near Kannauj.

Table 2.2. Rivers and river basins in the state of Uttar Pradesh

S.no.	Major basins		Prominent sub-basins	
	Name of river	Name of the major basin	Name of river	Name of sub-basin
1	2	3	4	5
1.	Yamuna	Yamuna Basin	Chambal	Chambal sub-basin
			Pahuj	Pahuj sub-basin
			Betwa	Betwa sub-basin
			Ken	Ken sub-basin
			Baghain	Baghain sub-basin
			Ohan	Ohan sub-basin
			Hindon	Hindon sub-basin
			Karona or Jhinjar	Karona or Jhinjar Sub-basin
			Khari	Khari sub-basin
			Sengar	Sengar sub-basin
			Arind	Arind sub-basin
			Tons	Tons sub-basin
2.	Ganga	Ganga Basin	Kali (east)	Kali (east) sub-basin
			Mahawa	Mahawa sub-basin
			Sot	Sot sub-basin
			Loni	Loni sub-basin
			Pandu	Pandu sub-basin
			Ishan	Ishan sub-basin
			Non	Non-sub-basin
			Varuna	Varuna sub-basin
			Karmnasha	Karmnasha sub-basin
			Belan	Belan sub-basin
3.	Ramganga	Ramganga basin	Khon	Khon sub-basin
			Koshi	Koshi sub-basin
			Khannaut	Khannaut sub-basin
			Garra	Garra sub-basin
			Sainjani	Sainjani sub-basin
			Aril	Aril sub-basin

4.	Gomti	Gomti basin	Sai Kalyani Bakulahi Behata Kathana	Sai sub-basin Kalyani sub-basin Bakulahi sub-basin Behat sub-basin Kathana sub-basin
5.	Ghaghra	Ghaghra basin	Sharda Suheli Sarju Little Gandak	Sharda sub-basin Suheli sub-basin Sarju sub-basin Little Gandak sub-basin
6.	Rapti	Rapti basin	Banganga Kuwano Ami Burhi Rapti Rohini	Banganga sub-basin Kuwano sub-basin Ami sub-basin Burhi Rapti sub-basin Rohini sub-basin
7.	Great Gandak	Great Gandak basin		
8.	Son	Son basin	Rihand Kanhar	Rihand sub-basin Kanhar sub-basin

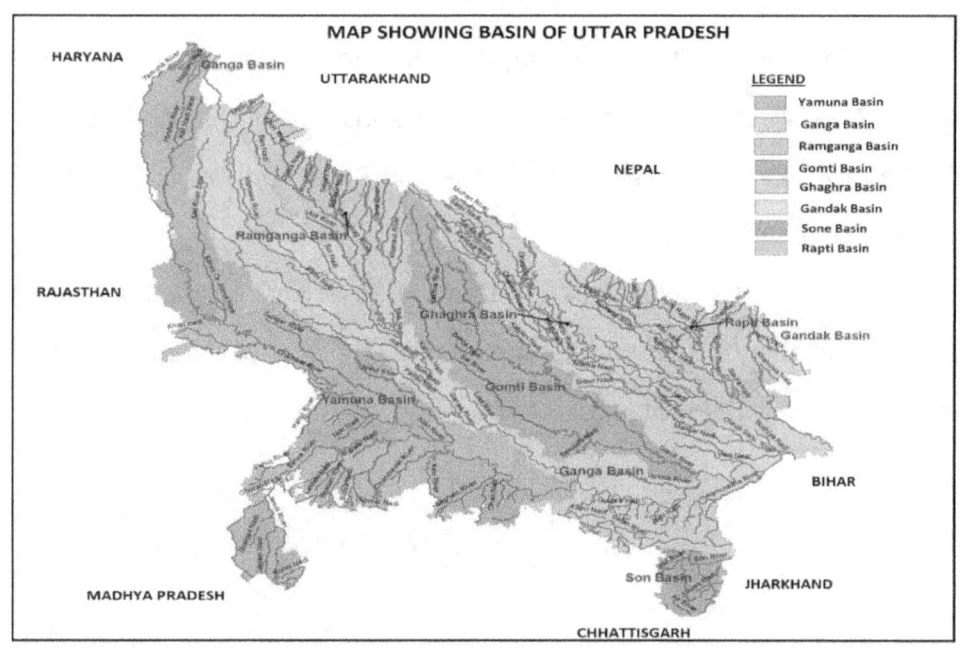

Fig. 2 Rivers and river basins in the state of Uttar Pradesh

Gomti River originates from a group of springs in the Tarai belt of the Pilibhit district. It flows southeasterly over a distance of 940 km before joining River Ganga at Gazipur.

Ghaghra River originates near Mansarover lake in China, enters Indian territory at Kanriyela. River Sharda joins it in district Sitapur. Saryu, Rapti, and Gandak are the left bank tributaries of the Ghaghra River. It covers a distance of 1100 km in the state.

Yamuna River originates from Yamnotri glaciers in the Himalayan range and after flowing a distance of 1400 km, it joins the Ganga River at Allahabad. Its main tributaries, Tons, Uma, Rishiganga, are in the hilly region, and Chambal, Betwa, and Ken originate from the southern plateau region.

The Chambal River emerges from Vindhyan ranges in Madhya Pradesh and, after flowing easterly, enters in U P at district Agra. The Betwa, another important right-bank tributary of Yamuna River, rises from the Vindhyan range in Madhya Pradesh, south of Bhopal, and flows northwards before it enters in U P at Lalitpur district. Finally, after flowing over a distance of 600 km, it joins the Yamuna River in the Hamirpur district. The River Tons rises in the Kaimur Range of Madhya Pradesh and, after flowing over a distance of 270 km, merges into Yamuna River in the Allahabad district.

The Son river neither rises in U.P. nor debouches in U.P. Rising in Madhya Pradesh, it flows through Sonbhadra district of U.P. and flowing a distance of 85 km it leaves U.P. and enters in Bihar and finally joins Ganga River near Danpur in Patna district [8].

There are 21 barrages, and 135 Dams & Reservoirs on various rivers in the state, and about 13 are under construction [7]. Some of the major Dams & Reservoirs are shown in Table 2.3.

Table 2.3 Major Dams and Reservoirs

Name of River	Name of Barrage/Dam
Ganga	Bhimgoda Barrage, Madhya Ganga Barrage, Narora Barrage, Kanpur Barrage
Yamuna	Hathnikund Barrage, Okhala Barrage, Gokul Barrage, Panchnada Dam
Hindon	Hindon Barrage
Kali East	Bewar Barrage
Ramganga	Kalagarh Dam, Harwali Barrage, Kho Barrage
Sharada	Banbasa Barrage, Sharada Barrage
Ghaghra	Girija Barrage
Banganga	Banganga Barrage
Narayani	Valmiki Barrage
Gomti	Gomti Barrage
Betwa	Rajghat Dam, Matatila Dam, Dhekwan Dam, Barua Dam, Paricha Dam
Shazad	Govind Sagar Dam, Shazad Dam
Sajnam	Sajnam Dam, Bhawani Dam
Jamini	Jamini Dam
Rohani	Rohini Dam
Arjun	Arjun Dam
Ken	Bariyapur Dam
Tons	Tons Barrage
Rihand	Rihand Barrage, Ghaghar Barrage, Obra Dam
Varuna	Varuna Barrage
Pahuj	Pahuj Dam, Dogri Dam

Magaria	Maudaha Dam
Urmil	Urmil Dam
Karmnasa	Naugarh Dam, Moosakhand Dam
Belan	Meja Dam

3. AQUIFER SYSTEM/ GROUNDWATER BASINS OF THE STATE

3.1 Occurrence of Groundwater

The occurrence of groundwater in the state is under varying situations. It occurs under water table or unconfined state down to a depth up to 50 to 100 m, and in deeper aquifers, it occurs in semi-confined to confined conditions in Ganga Plains. In Bundelkhand Plateau, occupied by hard rocks, it occurs in the secondary porosity under unconfined conditions. The water level cuts the ground level all along the northern border of the Tarai region in the form of the spring line. In the southern border of the Tarai Belt, groundwater occurs under high confinement in deeper aquifers, and flowing/artisan wells are conventional in localized patches.

The groundwater storage is largely controlled by the prevailing hydrological and geomorphic conditions, which are highly variable in the state. Rainfall is the primary source of recharge, which is highly variable in time and space. Abstraction is the main source of discharge which also varies in time and space. Apart from this, the magnitude of input (recharge) to the groundwater system and output (discharge) from it also influences the status of the groundwater regime [8].

3.2 Aquifer System

Uttar Pradesh is characterized by diverse hydrogeological setup, and various rock formations having different geological, geomorphological, and hydrological properties forming the various aquifer systems that comprise aquifers of different dimensions. These aquifers may be moderate to extensive both laterally and vertically in the alluvial region, and they may be unconfined, semi-confined to confined. In contrast, in the rocky terrain of Bundelkhand and Vindhyan regions, the aquifer systems are largely fissured, weathered, and discontinuous.

Without proper understanding and studying the dynamics and nature of the aquifers, any management action, if initiated, cannot succeed and this is one of the reasons for the mismanagement of groundwater resources in the state. A brief description of Aquifer Systems of the state is given in the following sub-sections [9].

3.2.1 Artesian Aquifer

Artesian aquifer belt in the state extends in front of the foothill region commencing from Maharajganj district in the east to near Saharanpur district in the west passing through Bijnor and Lakhimpur districts. This is a powerful aquifer system (auto flowing) that may provide water supplies for various uses, which do not need the energy to lift water from different depths. However, the recharge area of the artesian aquifer belt needs conservation and protection measures since the pressure 'heads' of the artesian aquifer have decreased with time [9].

3.2.2 Regionally Extensive Porous Aquifer

In UP, a multi-layered alluvial aquifer system having different depth ranges extends over Central Ganga Plain, which mostly covers rural areas. The aquifers are usually porous with varying granularity. Dimensions of aquifers may be regionally extensive, crossing administrative boundaries. Each aquifer group comprises a set of aquifers having different characteristics.

The sub-surface correlation of formations in the state has shown the presence of several aquifers down to a depth of 600 m or beyond below the ground level, broadly comprising 4 Aquifer Groups at different depth ranges, which is variable in different parts of the state within Alluvial Plain. These aquifers, mainly encountered in Central Ganga Plain have been grouped based on lithological characters as well as based on interpretation of electrical logs of drilled boreholes, which are summarized in Table 3.1[3].

Table 3.1.Different aquifer groups in Central Ganga Plain

S.no.	Description	Depth range below ground level in m	
1.	First aquifer	0.0	150.00
2.	Second aquifer	100.00	210.00
3.	Third aquifer	225.00	360.00
4.	Fourth aquifer	360.00	600.00 or beyond

The granularity of these aquifers controls the yield and flows of groundwater. Coarser the grain size of aquifer material, the higher the yield as well as, the storage. The upper part of the first aquifer down to 50 m below ground level (bgl) is the primary source of water supply through hand pumps, dug wells, and shallow tube wells and is unconfined. It is also called a shallow or phreatic aquifer, and it is under heavy stress. The first aquifer as a whole, which is under unconfined to semi-confined conditions, is the most potential aquifer group, which is the main source of groundwater in the state extensively exploited through private as well as government tube wells to meet the drinking water, irrigation, and other needs.

The second aquifer group is mostly dominated by saline water, covering a larger part of Central Ganga Plain. However, in the northern fringe of this central Ganga Plain, salinity is not reported. The granularity of aquifers is fine with low to moderate yield. The third and fourth aquifer groups are also characterized by fine sand, and the yield is low to moderate. The aquifers are mostly semi-confined or confined in nature.

The depth ranges of all the aquifer groups are highly variable across the state. The minimum and maximum depth of these aquifer groups of Central Ganga Plain may vary from west to east and north to south. The granularity of aquifers is also quite variable. Towards Agra, Mathura, Firozabad, usually third and fourth aquifer groups are not encountered, and the bedrock may be encountered at a depth of 150-200 m bgl. Therefore, for the planning purpose, an area-specific aquifer system should be taken into consideration. The deeper aquifers are being exploited by the private as well as government tube wells to meet the drinking water and irrigation needs.

3.2.3 Urban Aquifer System

Though various major and small towns form part of the Central Ganga Plain Aquifer system, it needs special mention as the development potential and recharge techniques differ from those with rural area aquifers. The urban aquifer illustrations so far prepared for some urban areas depict in general, a multi-layered aquifer system distinguishable with four aquifer groups at different depths and covered with clay layers up to 20 m thick. Following urban aquifer groups are mostly encountered in the state [9]:-

(a)	First aquifer group	-	within 100 m depth.
(b)	Second aquifer group	-	between 130 m & 255m depth
(c)	Third aquifer group	-	between 200 m & 460 m depth
(d)	Fourth aquifer group	-	between 500 m & 753 m depth

However, the above depth ranges vary from area to area. Therefore, different cities are characterized by different aquifer systems of variable depth ranges. The aquifers beyond 100 m are in varying degrees of confinement (semi-confined to confine) and yield a large quantity of groundwater ranging from 1000 to 3000 liters per minute. The first aquifer is already under stress, and it has been observed that tube well yields are continuously reducing. Therefore, this aquifer needs its sustainable yield level to be fixed for the longevity of the aquifer for which groundwater mapping and modeling of urban aquifers are needed on priority.

3.2.4 Discontinuous Aquifer System in Rocky terrain

The southern peninsular region of Bundelkhand and Vindhyan comprises discontinuous aquifers of limited resource potential in a weathered and fissured/fractured system. The weathered mantle, overlying the rocks, envisages phreatic aquifers of non-confining nature. The fissured aquifers comprising fissures, fractures, cracks, joints within the rock formations, contain groundwater under a semi-confining state, and the storage in such aquifers is low to moderate, depending upon the thickness and extent of the fissured/weathered zone. Here the network of fissures and fractures serves as a permeable conduit feeding water to the wells. The flow of such groundwater remains limited within the watershed unit.

3.2.5 New Setup of Aquifer Systems of Uttar Pradesh

The above aquifer systems are usually quoted and used in various groundwater studies and deliberations. In 2012, the Central Ground Water Board (CGWB) further classified and rearranged the above aquifer systems in a more focused manner. The new setup of the aquifer system at the National level has been classified into 14 Principal Aquifer Systems, which are further sub-classified into 42 Major Aquifers. For this classification, geological and geomorphological formations with distinctive hydrogeological characteristics and parameters have been considered. Based on the above classification, the diverse hydrogeological setup of Uttar Pradesh is broadly grouped into the following principal aquifer systems [10].

3.2.5.1 Alluvial Aquifers

This Aquifer system covers a major part of UP. The general hydrogeological characteristics have already been discussed for the aquifer system of the Ganga Plain of the state. Two major aquifers under this aquifer system are identified-

Older Alluvial Aquifer: It covers a major part of the state comprising older alluvium sediments. The CGWB has recommended a specific yield for this aquifer as 6 %, with a range of 4 % to 8 %. This is an essential parameter for groundwater resource assessment.

Younger Alluvial Aquifer: It usually extends over present-day flood plain area and comprises newer alluvium material. The specific yield for this aquifer is recommended as 10 %, with a range of 8 to 12 %.

The specific yield denotes typically the volume of water that a groundwater system releases, and it represents the water yielding capacity of saturated sediment. It is defined as the ratio of the volume of water that a saturated soil or rock will yield by gravity to the total volume of soil or rock. Specific yield is usually expressed as a percentage and it is also known as effective porosity.

3.2.5.2 Principal Aquifer Systems in Bundelkhand and Vindhyan Region

Based on the geological formations and occurrence of rock types, five broad principal aquifer systems identified for Bundelkhand and Vindhyans are:-

(i)   Sandstone aquifer
(ii)   Shale aquifer
(iii)   Limestone aquifer
(iv)   Granite aquifer
(v)   Gneissic aquifer

These principal aquifers comprise different major aquifers, as coded by the CGWB for Southern Peninsular UP. These major aquifers are mostly discontinuous and localized and vary from place to place. The prevailing groundwater conditions are already discussed above. The maps of these major aquifers are available with the Northern Regional office, CGWB.

4. GROUNDWATER AND SURFACE WATER MONITORING IN THE STATE

4.1 Surface Water Monitoring

Over the years, a robust monitoring network has been developed for surface water in the state. The regional office of the Central Water Commission (CWC) at Lucknow is responsible for hydrological observation in the state. CWC at present operates a national network of 954 hydrological observation stations to collect: (I) water level, (II) discharge, (III) water quality, (IV silt, and (V) selected meteorological parameters including snow observation at key stations. CWC also maintains a three-tier laboratory system to analyze the quality of water in rivers. Parameters like temperature, color, odor, specific conductivity, total dissolved solids, PH, dissolved oxygen, physical and chemical characteristics, biochemical oxygen demand, traces of toxic elements, polyaromatic hydrocarbon, insecticides, and micro in the water are also analyzed. CWC also monitors reservoir levels and live storage capacity in respect of 63 important reservoirs in India [11, 12]. More details of the surface water network cannot be revealed further as the state of UP lies in Ganga Basin, which is a classified basin in India.

State also started to modernize and automatize information and data acquisition setup. Progress of installed Hydrological Information Sensors in U P is shown in Table 4.1[7].

Table 4.1 Installed Hydrological Information Sensors

Description	Installed by Irrigation and Water Resource Department	Installed by Central Water Commission	Installed by Indian Metrological Department
Automatic Weather Stations	26	-	50
Automatic Rain Gauges	-	-	125
Rain Gauges	71	58	-

4.2 Groundwater Monitoring

Both the central and state agencies monitor groundwater in the state. The northern region office of CGWB, based at Lucknow, monitors about 1198 Groundwater Water Monitoring Wells that include 900 dug wells, and 298 purpose-built piezometers establish in the state of UP. Apart from this participatory monitoring program for the remaining 8 months is also conducted by the board. All stations are monitored four times in a given hydrological year in August (20th to 30th day), November (post-monsoon) (1st to 10th day), January (1st to 10th day), and May (pre-monsoon) (20th to 30th day). The long-term data generated during these monitoring seasons are used for computation, comparison, and analysis of groundwater utilization and its availability [8].

Ground Water Department of the state of UP also monitors groundwater through a massive network of about 9000 hydrograph stations (observation wells and piezometers) both in rural and urban areas. In rural areas, these hydrograph stations are located in a grid of 5 km x 5 km, and in urban areas, in a grid of 2 km x 2 km. As a standard practice, groundwater levels in rural areas are measured four times a year that includes pre-monsoon and post-monsoon readings, one reading in August during the monsoon period and the fourth reading is taken in February/March. In urban areas, monthly groundwater level measurements are taken. The water level monitoring is done manually through handheld steel tapes or sounders. Such data is primarily used to assess the periodic water level fluctuation and trend analysis for the resource assessment.

The groundwater conditions have a direct impact on groundwater levels; therefore, groundwater level measurements from observation wells/piezometers are the principal source of information about the seasonal fluctuation and the hydrologic stresses acting on aquifers. These stresses with groundwater level changes affect groundwater recharge, storage, and discharge within a hydrological regime.

4.2.1 Real-Time Measurements

In dynamically changing geo environment impacted with widespread groundwater depletion, timely monitoring of groundwater levels has become a critical issue, and therefore, the necessity of groundwater level measurements for gathering real-time data through Digital Water Level Recorder (DWLR), using telemetry system, have come to the forefront [13].

Under the World Bank-funded Uttar Pradesh Water Sector Restructuring Project (Phase-2), a separate Groundwater Component was approved for the state of UP with the prime objective to strengthen the existing groundwater monitoring network, to establish a data storage system with analysis and management, and to develop aquifer and conjunctive water use management plan along with capacity building and institutional development of Ground Water Department.

Under the project, it is proposed to install 1050 DWLRs with Telemetry on existing piezometers located in different river sub-basins, covering part of Yamuna, Gomti, Ganga, Ramganga basins. The installation of DWLRs started in the year 2015 in a phased manner. A total of 1050 DWLRs have been installed in Division Agra, Aligarh, Lucknow, Kanpur, Allahabad, Varanasi, Moradabad, Saharanpur, Meerut, Jhansi, and Chitrakoot at selected locations in the state. Groundwater level data are received daily at an interval of 12 hours (6 am & 6 pm) [14].

Apart from this, under the National Hydrology Project, key piezometers with digital water level recorders have recently been installed in 149 problematic blocks and 22 stressed cities of the state.

4.2.2 Multiple Monitoring Stations

Under the World Bank Project UPWSRP-2, 100 Multiple Monitoring Stations have been installed in the state. Such multiple monitoring stations are designed to have 3 piezometers of varying depths; deep (up to 250 m), medium (up to 150 m), and shallow (up to 80 m) to tap the different aquifers. Presently, groundwater level measurements are done on shallow/dynamic aquifers through shallow monitoring wells. Though the groundwater extraction has already started from deeper aquifers, the impact of extraction on the changes in their piezometric levels could not be observed so far, as the piezometers of deeper depths were not installed previously.

With the help of multiple monitoring wells, the changes in groundwater level fluctuation for varying depths will be easily observed. This information will also be useful in determining the aquifer characteristics as well as in the assessment of in-storage/static groundwater resources. These multiple monitoring stations are equipped with DWLRs for real-time monitoring of groundwater levels providing daily observation on the piezometric levels in the piezometers of different depths [14].

However, the Ground Water Department, UP, lacks the necessary infrastructure and capacity to analyze daily data and use it in the decision-making process as a Decision Support System (DSS) is not in place. Without DSS there is no utility of real-time monitoring.

5. DEVELOPMENT OF IRRIGATION SOURCES

UP is one of the states in India where tremendous water development has taken place after independence for extension of irrigation facilities and to fulfill other requirements. In the eighties, the state started many schemes to harness groundwater for irrigation, which, coupled with the advent of irrigation pump sets, ignited the tube well revolution, and Uttar Pradesh became the center of the tube well revolution in India. The exponential increase was observed in the installation of tube wells in the state as groundwater was readily available with comparatively less investment, easy to abstract in the state, and groundwater irrigation source is an assured source of water supply. Canal network also grew, though marginally. The progress of different irrigation sources is shown in Table 5.1 [15, 16].

Table 5.1   Cumulative progress of different irrigation sources

S.no.	Year	Canal (km)	Dug wells (nos.)	Private tube wells (nos.)	State tube wells (nos.)
1	2	3	4	5	6
1.	1984-85	66262	717485	1885275	22294
2.	2002-03	68869	125073	3664549	28527
3.	2004-05	71782	125441	3774453	28124
4.	2006-07	NA	126000	3909000	NA
5.	2011-12	73599	129000	4160000	NA
6.	2013-14	74127	139000	4229000	30917
7.	2014-15	74882	140000	4251100	32047
8.	2016-17	75043	148000	4295000	33375
9.	2017-18	NA	155000	4318600	33351

Note: NA – Not available

From Table 5.1, it is clear that after the year 1984–85, the number of private tube wells increased exponentially in the state from 1.88 million in the year 1984-85 to 4.318 million in the year 2017-18 or an increase of more than 300 %. Canal network also increased during this period, though marginally, from 66,262 kilometers to 75,043 kilometers or an increase of only 13 %. A sharp decline in the number of open wells or dug wells is observed between the year 1984–85 to the year 2002 - 2003 from 0.717 million to 0.125 million. After that, it increased continuously to 0.155 million in 2017–18. This is mainly due to the implementation of the Community Dug/Blast Well Scheme under the Rural Infrastructure Development Fund (RIDF) of the National Bank for Agriculture and Rural Development (NBARD) in Bundelkhand and Vindhyan Regions having rocky and semi rocky formations. Dug cum blast wells constructed under Bundelkhand Package also contributed to it.

The above table also shows that private tube wells are the primary source of irrigation in the state. But these tube wells have a short life span as a majority of them are shallow tube wells having a life span of 10 to 15 years. To assess the actual number of minor irrigation works in use, a census of minor irrigation works is conducted in the state at regular intervals under centrally sponsored Rationalization of Minor Irrigation Statistics Scheme, and based on actual figures; the number of minor irrigation works is rationalized. Data shown in Table 6.1 do not incorporate the minor irrigation census figures after 2001. After 2001, the Fourth and Fifth Minor Irrigation (MI) Census having the reference year 2006-07 and 2013-14 respectively had been conducted. Based on the Fifth Census of Minor Irrigation, the status of minor irrigation schemes is shown in Table 5.2 [17, 18].

Fifth M I Census data show that dug wells are the primary minor irrigation source in the country followed by shallow tube wells, while in the state of UP, shallow tube well are the main minor irrigation source followed by medium-deep tube wells, deep tube well and dug wells. Fifth M I Census data also show that out of about 21.7 million minor irrigation schemes in the country, about 3.8 million or approximately 17.5 % are in the state of U P. Out of 0.87 million dug wells in the country 0.108 million, or about 12.4 % are in the state of UP. Out of 5.94 million shallow tube wells in the country, 3.33 million, or about 56.06 % are in the state of UP. Out of 3.17 million medium-deep tube wells in the country, 0.25 million, or about 7.8 % are in the state of UP. Out of 2.618 million deep tube wells in the country, 0.087 million, or about 3.32 % are in the state of UP.

Table 5.2   Minor irrigation structures

S. no.	Minor irrigation source	As per the Fifth Census of Minor Irrigation Schemes (the reference Year 2013-14			
		In India		In Uttar Pradesh	
		Nos. (million)	%	Nos. (million)	%
1	2	3	4	5	6
1.	Dug Wells	8.785	40.66	0.1086	2.86
2.	Shallow Tube Wells	5.940	27.36	3.3323	87.66
3.	Medium Deep Tube Wells	3.176	14.63	0.2501	6.58
4.	Deep Tube Wells	2.6187	12.06	0.0872	2.3
5.	Surface Flow Schemes	0.5921	0.27	0.0081	0.21
6.	Surface Lift Schemes	0.6000	0.28	0.0147	0.39
	Total	21.714		3.8012	

In MI Census all the irrigation structures which are in use having cultural command area up to 2000 hectares are categorized as Minor Irrigation Schemes and enumerated through a village-level survey. Schemes having cultivable command areas between 2000 hectares to 10000 hectares are classified as Medium Irrigation Schemes and above 10000 hectares as Major Irrigation Schemes. Tube wells up to the depth of 30 m are classified as shallow. Tube wells between the depths of 30 m to 70 m are considered as medium-deep, and above 70 m are considered as deep. MI Census also includes state tube wells, which are constructed, owned, and operated by government departments.

If figures of the 5th MI Census are compared with the figures of the year 2013-14 in Table 5.1, a huge difference is seen in the numbers of tube wells. According to Table 5.1 total number of tube wells (private + state) comes out to be 4.2459 million while as per the 5th MI Census it comes out to be 3.669 million (shallow tube wells + medium-deep tube wells + deep tube wells), meaning thereby that about 0.5769 million tube wells became defunct over the time and if the figure of 3.669 million is considered as the total number of tube wells in the year 2013-14 as per 5th MI Census, the realistic figure of 2017-18 would be about 3.7586 million.

One of the major contributors for the construction of a large number of shallow tube wells in the state of UP is the Free Boring Scheme launched by the state government, specifically, for small and marginal farmers in the year 1984-85 for the construction of shallow tube wells as the percentage of small and marginal farmers is quite high, about 90 % in the state. This scheme had been the flagship scheme of the state as far as the creation of irrigation facilities is concerned and changed the lives of millions of farmers in the state by providing assured sources of irrigation at a very low cost. The number of bore wells constructed year-wise under the scheme and increase in net irrigated area is shown in Table 5.3[19].

Table 5.3 Cumulative progress of shallow boring and irrigated area

Year	Shallow boring (Nos.)	Net irrigated area (lakh hectare)
1984-1985	5481	101.54
1985-1986	59927	101.55
1986-1987	120434	98.54
1987-1988	185055	100.43
1988-1989	354317	101.69
1989-1990	540482	103.32
1990-1991	739062	105.42
1991-1992	937854	110.48
1992-1993	1130039	113.22
1993-1994	1324136	115.64
1994-1995	1598881	116.70
1995-1996	1884001	117.47
1996-1997	2168832	119.99
1997-1998	2203184	120.12
1998-1999	2304435	126.91

1999-2000	2344470	124.70
2000-2001	2420242	124.01
2001-2002	2501999	128.28
2002-2003	2576612	128.48
2003-2004	2664527	132.27
2004-2005	2804484	131.19
2005-2006	3011232	130.75
2006-2007	3232364	133.13
2007-2008	3491171	130.85
2008-2009	3652160	134.35
2009-2010	3739192	133.83
2010-2011	3910575	134.40
2011-2012	4084359	138.09
2012-2013	4220600	139.29
2013-2014	4312700	140.27

(Unit: 10 Lakh Hectare = 1 Million Hectare)

Table 5.3 suggests that since its inception up to the year 2013-14, 4.3 million bore wells have been constructed under Free Boring Scheme in the state and this scheme has played a pivotal role in increasing irrigated areas in the state.

Sources of energy being used in M I Schemes are shown in Table 5.4 which reveals that at the national level electricity is the main source of energy followed by diesel and other sources which include wind, solar, manual/ animal, etc. However, in the state of Uttar Pradesh diesel is the main source of energy followed by electricity and other sources [18].

Table 5.4 Minor Irrigation Schemes according to the source of Energy as per 5th MI Census

Source of energy used in MI schemes	In UP		In India	
	Number of schemes	% of the total	Number of schemes	% of the total
Electric	515976	14.8	13723254	72.77
Diesel	2964354	85.05	4768259	25.28
Others	5352	0.15	366943	1.95
Total	3485682		18858456	

6. IRRIGATED AREA AND CROP PRODUCTION

6.1 Irrigated Area

6.1.1 Definitions

1) Gross Command Area (GCA) – It is the total area that can be irrigated by a particular channel or project. It includes the area covered by roads, culverts, uncultivable areas, villages, etc. [20].

2) Cultural Command Area (CCA) – It is the portion of GCA, which is cultural or cultivable. The CCA is equal to GCA minus the uncultivable area in GCA [20].

3) Net Irrigated Area – It is the maximum area irrigated in one crop season out of CCA during a reference year.

4) Gross Irrigated Area – The area under various crops during a year, counting the area irrigated under more than one crop during the same year as many times as the number of crops grown and irrigated [21].

5) Irrigation Potential Created (IPC) – The total gross area proposed to be irrigated under different crops during a year by a scheme. The area proposed to be irrigated under more than one crop during the same year is counted as many times as the number of crops grown and irrigated [21].

6) Irrigation Potential Utilized (IPU) – Gross area irrigated during reference year out of the gross proposed area to be irrigated during a year [21].

6.1.2 Source Wise Irrigated Area

At present, the state boasts of having about 87 % irrigated area while the national average is only about 49 %. Before the five-year plan period, the irrigation potential created was 5.4 million hectares only, which is now more than 36 million hectares, though there is a huge gap between potential created and potential utilized. The net irrigated area also increased from about 3.2 million hectares to approximately 14.4 million hectares after independence. Thus, the state is now categorized as one of the highly irrigated states of the country and behind the state of Punjab only in terms of irrigated areas [16, 22]. Source-wise irrigated area over the years is presented in Table 6.1[15, 16].

Table 6.1 Source wise irrigated area in the state of Uttar Pradesh

S.no.	Name of source	Net irrigated area (million hectares)				
		1984-85	2002-03	2006-07	2011-12	2014-15
1	2	3	4	5	6	7
1.	Canals	3.331	2.630	2.614	2.555	2.482
2.	State tube wells	0.654	0.505	0.373	0.490	0.444
3.	Private tube wells	5.086	8.783	9.129	9.671	9.739
4.	Other sources	1.083	0.928	1.197	1.093	1.724
	Total	10.154	12.846	13.313	13.809	14.389

Table 6.1 suggests that net irrigated area through canal and state /government tube wells has a declining trend, while net irrigated area through private tube wells and other sources has an increasing trend. The irrigated area through private tube wells has almost doubled between the years 1984-85 to 2014-15. The irrigated area through other sources also shows an increasing trend. In other sources, open dug wells are the main source of irrigation as shown in Table 6.2[16].

Table 6.2 The net irrigated area through other sources

S. no.	Name of source	Net irrigated area (million hectares)			
		2006-07	2011-12	2012-13	2014-15
1	2	3	4	5	6
1.	Dug wells	0.986	0.848	1.308	1.473
2.	Tanks, lakes and ponds	0.149	0.099	0.119	0.186
3.	Others	0.062	0.146	0.060	0.067
	Total	1.197	1.093	1.487	1.724

Table 6.2 shows that net irrigated areas through wells and tanks, lakes, and ponds also show an increasing trend. The net irrigated area from these sources almost increased 1.5 times between the year 2006-07 and 2014-15.

Net area irrigated through surface water and groundwater is presented in Table 6.3[16]. Table 6.3 suggests that 81 % of the net irrigated area is attributed to groundwater resources, while about 19 % of the net irrigated area is attributed to the surface water resource. Thus irrigated agriculture in the state of Uttar Pradesh is heavily dependent on groundwater resources.

Table 6.3 The net irrigated area through Surface and Groundwater Resources (2014-15)

S.no.	Resource type	Net irrigated area (million hectares)	Percentage of total net irrigated area
1.	Surface water		
	a. Canal	2.482	17.249
	b. Tanks, lakes/ponds	0.184	1.27
	c. Others	0.067	0.96
	Subtotal	2.733	18.99
2.	Groundwater		
	a. State tube wells	0.444	3.08
	b. Private tube wells	9.739	67.68
	c. Dug wells	1.473	10.23
	Subtotal	11.656	81.00
	Total	14.389	

6.1.3 Coverage with Modern Irrigation Methods

Though the state did well in increasing the irrigated area, the adoption of modern irrigation systems is comparatively sluggish in the state. Data obtained in the 5th MI Census are presented in Table 6.4[18].

Table 6.4 indicates that the adoption of water-efficient micro irrigation systems is only 0.25 percent in the state of UP while at the national level, this percentage is 5.2. A large number of farmers still use open channels to irrigate their fields. A positive feature is that the use of surface pipe is increasing and its percentage is quite high at the national level as well as in the state of UP also.

Table 6.4 Number of minor irrigation schemes with the type of water distribution method

S.no.	Water distribution method	5th M I Census (the reference year 2013-14)			
		In India		In Uttar Pradesh	
		Nos. (million)	% of total	Nos. (million)	% of total
1	2	3	4	5	6
1.	Open water channel	11.0522	52.97	2.4848	66.13
2.	Underground pipe	3.1261	14.98	0.0336	0.90
3.	Surface pipe and others	5.6184	26.92	1.2312	32.71
4.	Micro-irrigation				
	i)Drip	0.3880	1.86	0.0039	0.10
	ii)Sprinkler	0.6813	3.27	0.0054	0.15
	Sub total	1.0693	5.12	0.0093	0.25
	Total	20.8662		3.7589	

The astonishing feature is that despite a high level of subsidy for sprinkler and drip irrigation, the adoption of micro-irrigation is still very low at the national level and almost negligible in the state of UP. Thus Micro Irrigation programs need a relook. It should be designed according to the specific conditions of a particular state.

6.1.4 Creation and Utilization of Irrigation Potential

With an increase in irrigation facilities, irrigation potential also increased substantially in the state. Table 6.5 shows the growth of irrigation potential and its utilization in the state of UP [16, 23].

Table 6.5 suggests that there is a huge gap between potential created and utilized, which has not improved over time. In the year 1996-97 created irrigation potential was 28.725 million hectares, and its utilization was 20.547 million hectares or 71.5 % only. In the year 2015-16, the created irrigation potential was 36.7 million hectares, and its utilization was 24.69 million hectares or 67.27 %.

Table 6.5 Growth of irrigation potential and its utilization

Year	Irrigation Potential Created (million hectares)				Irrigation Potential Utilization (million hectares)			
	Minor Irrigation		Major & Medium Irrigation	Total	Minor Irrigation		Major & Medium Irrigation	Total
	State	Private			State	Private		
1	2	3	4	5	6	7	8	9
1996-97	3.83	17.85	7.04	28.72	1.86	12.57	6.11	20.55
1997-98	3.85	18.16	7.17	29.18	1.87	12.25	6.15	20.269
1998-99	3.87	18.5	7.28	29.67	1.88	11.99	6.22	20.10
1999-2000	3.87	18.66	7.41	29.94	1.88	11.54	6.29	19.721
2011-12	3.95	23.06	8.53	35.55	1.87	14.61	7.20	23.71
2013-14	3.99	23.64	8.78	36.42	1.91	15.07	7.42	24.40
2014-15	4.04	23.71	8.84	36.68	1.93	15.13	7.50	24.56
2015-16	4.071	23.77	8.85	36.70	1.93	15.17	7.58	24.69

So, as far as utilization of irrigation potential is concerned, no significant improvement is seen in 19 years, rather it deteriorated somewhat. Problem areas are private minor irrigation and state minor irrigation in which potential utilization was 63.81 % and 47.4 % respectively in the year 2015-16. Probable reasons between the gap in potential created and utilization may be the following:-

(I) There may be over-reporting of created irrigation potential as it has surpassed the assessed ultimate irrigation potential. The assessed ultimate irrigation potential of the state including Uttarakhand is 25.7 million hectares of which, 12.5 million hectares are attributed to Major& Medium irrigation projects and 13.2 million hectares are attributed to Minor irrigation projects

[11]. Potential created through State and Private minor irrigation in the year 2015-16 is 27.82 (4.071+23.75) million hectares, which is far more than the assessed ultimate irrigation potential of 13.2 million hectares.

(II) Under minor irrigation, the majority of schemes are ground water-based. Ground water-based schemes have a shorter life span, especially private minor irrigation schemes, and become defunct in due course of time. To assess the actual situation, a census of minor irrigation works is conducted in the country at regular intervals. As per the 5[th,] MI Census (the reference year 2013-14) potential created and potential utilized are shown in Table 6.6[18].

Table 6.6 Irrigation potential created and utilized through minor irrigation schemes as per the Fifth Census of Minor Irrigation Scheme (The reference year 2013-2014)

In million hectares

Type of minor irrigation Schemes	In India			In the State of Uttar Pradesh		
	IPC	IPU	%	IPC	IPU	%
1	2	3	4	5	6	7
Ground water Schemes	77.64	63.06	81.20	22.54	16.70	74.10
Surface water Schemes	9.689	7.779	80.26	0.1227	0.0909	74.06
Total	87.33	70.82	81.09	22.66	16.79	74.09

Note: IPC – Irrigation Potential Created, IPU – Irrigation Potential Utilized

According to Table 6.5 potential created in the year, 3013-14 through minor irrigation schemes is 27.63 (3.99+23.64) million hectares while, as per Table 6.6 it was found as 22.66 million hectares. Similarly, according to Table 6.5 potential utilized in the year 2013-14 was 16.981 (1.914+15.067) million hectares while, according to Table 6.6, it was found as 16.79 million hectares. Thus there is a huge gap between the figures of created irrigation potential while figures of potential utilization are almost the same indicating that the figures of created irrigation potential have not been reconciled with MI census figures and do not include loss of irrigation potential over time.

(III) For private minor irrigation works, the potential created is calculated based on certain norms fixed for each work. For example, the norm of irrigation potential for shallow tube well is 05 hectares, which has not been revised for many years. With a substantial increase in the number of shallow tube wells over the years, the irrigation potential of shallow tube wells may have been reduced. So there is a need to have a relook at these norms and revise them if required. This may lead to revising the design of present tube wells, which may result in reducing the cost as the reduced norm may reduce the diameter of the tube well.

6.1.5 Gross Irrigated Area and Irrigation Potential Utilized

A comparison of gross irrigated area and irrigation potential utilized is presented in Table 6.7[16, 23].

Table 6.7 Comparison of gross irrigated area and irrigation potential utilized in the State of UP

million hectares

S.no	Year	Gross irrigated area through all sources	Irrigation potential Utilized through all sources
1	2	3	4
1.	1998-99	17.698	20.097
2.	2011-12	19.901	23.710
3.	2013-14	20.403	24.403
4.	2014-15	20.965	24.560

Gross irrigated area and irrigation potential utilized vary year to year. In an ideal situation, gross irrigated area and irrigation potential utilized should be the same for a particular year. But Table 6.7 indicates that there is a vast difference in these figures for a specific year and there is a necessity to rationalize the statistics.

6.1.6 Water Logging

Areas of canal command face problems of waterlogging and land degradation in the state. Waterlogging is a hydrological condition when the groundwater table reaches shallow, critical, and semi-critical conditions in which the root zone and soil profiles are fully or partially saturated with capillary groundwater. For different soil groups and different crops/ trees, the critical and semi-critical water level is different, because of different limits of capillary rise of groundwater and depth of the root zone. Considering the ecological and agricultural viewpoints waterlogged areas can be grouped into four classes [24]:-

Ponded/ flooding conditions

This is the condition where water gets accumulated on the surface temporarily or for a longer duration and in most cases groundwater reaches above the ground. Such water accumulations are common in the depression zones and paleochannels and get increased due to the escaping water of canal systems.

Marshy lands

In this class fall those waterlogged areas where, during most of the year, the groundwater level is above the ground surface and less than 0.5 meters below ground level, during part of the year. Marsh weeds are its distinguishing feature.

Critical waterlogged areas

These are those areas where the groundwater level is in critical condition, and groundwater reaches the surface with capillary rise. For the different soil groups, the critical water depth is different, varying from 1.5 to 3 meters below the ground level.

Semi critical waterlogged areas

Such areas during part of the year, especially in monsoon and post-monsoon months are in critical zones and during pre-monsoon months in the semi-critical zone, where the capillary rise of groundwater during most of the year reaches the root zones but not always to the surface.

Systematic mapping and delineation of waterlogged areas have not been done in the state, though; several attempts have been made from time to time to assess the waterlogged areas. Some of them are described below:–

a) National Commission on Agriculture assessed the extent of waterlogging in the country in the year 1976 according to which, 0.810 million hectares were waterlogged in the state of UP. The Ministry of Agriculture, Government of India, estimated in 1984-85 that an area of 1.98 million hectares was suffering from the problem of waterlogging in the state of UP, including irrigated and unirrigated areas. The working group constituted by the Ministry of Water Resources, Government of India in the year 1991 assessed that an area of 0.430 million hectares was suffering from the problem of waterlogging under irrigation commands in the state of UP [25].

b) Regional Remote Sensing Service Center (RRSSC), Indian Space Research Organization, Jodhpur, Government of India and Central Water Commission (CWC), Government of India, in its report "Assessment of Water Logging and Salt and/or Alkaline Affected Soils in the Commands of all Major and Medium Irrigation Projects in the Country Using Satellite Remote Sensing", gave various estimates of waterlogged areas in the country during the years 2003- 05.

c) The said report estimated perennial and seasonal waterlogged areas as shown in Table 6.8 [26]. Table 6.8 shows, that only 0.54 % area of the canal commands was waterlogged during the years 2003-05 in the state of U P. But the said report also estimated critical areas of depth of groundwater for pre-monsoon and post-monsoon during 2003-05, and these estimates are shown in Tables 6.9 and 6.10 [26]. Table 6.9 suggests that during the years 2003-05, critical areas according to the depth of groundwater was only 2.6 %, while at all India level it was 13.86 %. So comparatively, U P was in a far better position. Table 10 indicates that in the post-monsoon season situation in U P comparatively worsened.

Table 6.8 Waterlogged area in the commands of Major & Medium Irrigation Projects as assessed by RRSSC

million hectares

Name of the assessed unit	Geographical area	Command area	Waterlogged area			
			Perennial	Seasonal	Total	% of Command area
1	2	3	4	5	6	7
Uttar Pradesh	24.092	23.4007	0.01129	0.11538	0.127	0.54
India	328.72	88.895	0.1731	1.54613	1.719	1.93

Table 6.9 Critical areas of depth of groundwater during 2003-05 for pre-monsoon (within commands)

Category	Uttar Pradesh	India
Most Critical < l m		
million hectares	00.00	0.1250
% of Command area	00.00	0.14
Critical 1-2 m		
million hectares	0.00584	2.4456
% of Command area	0.02	2.75
Less Critical 2-3 m		
million hectares	0.6081	9.750
% of Command area	2.6	10.97
Non Critical > 3 m		
million hectares	22.786	76.574
% of Command area	97.34	86.14

(d) Ground Water Department UP (GWDUP) and Central Ground Water Board (CGWB) estimated waterlogged areas, measuring groundwater level on selected hydrograph stations. It was recommended that the period of waterlogging estimation might be different for different crops. Keeping in view the critical Rabi crops, the water level of October/November (post-monsoon) should be considered for waterlogging and presented estimates of waterlogged areas in the state of UP, as shown in Table 6.11[24] which shows that waterlogged area in UP was 6.344 million hectares in the year 2004.

Table 6.10 Critical areas of depth of groundwater for Post Monsoon (within commands) during 2003-05

Category	Uttar Pradesh	India
Most Critical < l m		
million hectares	00.02959	1.7368
% of Command area	00.13	1.95
Critical 1-2 m		
million hectares	0.9193	10.979
% of Command area	3.93	12.35
Less Critical 2-3 m		
million hectares	4.092581	18.186
% of Command area	17.49	20.46
Non Critical > 3		
million hectares	18.359	57.993
% of Command area	78.45	65.24

Table 6.11 The waterlogged area over the years in post-monsoon November month in the state of U.P. as estimated by GWDUP & CGWB

In million hectares

S.no	Zones							
	Eastern Plains		Central Plains		Western Plains		State	
1	2		3		4		5	
1	Reported area		Reported area		Reported area		Reported area	
	Year	Area	Year	Area	Year	Area	Year	Area
	1999	7.502	1999	4.628	1999	8.077	1999	20.107
	2000	7.566	2000	4.583	2000	8.081	2000	20.230
2.	Critical waterlogged area (groundwater level 0 to 2 m)		Critical waterlogged area (groundwater level 0 to 2 m)		Critical waterlogged area (groundwater level 0 to 2 m)		Critical waterlogged area (groundwater level 0 to 2 m)	
	Year	Area	Year	Area	Year	Area	Year	Area

	1996 2000 2004	2.359 1.570 1.017	1996 2000 2004	1.135 0.726 1.004	1996 2000 2004	1.090 0.981 0.685	1996 2000 2004	4.584 3.277 2.706
3.	Semi critical waterlogged area (groundwater level 2 to 3 m)		Semi critical waterlogged area (groundwater level 2 to 3 m)		Semi critical waterlogged area (groundwater level 2 to 3 m)		Semi critical waterlogged area (groundwater level 2 to 3 m)	
	Year	Area	Year	Area	Year	Area	Year	Area
	1996 2000 2004	1.731 2.512 1.736	1996 2000 2004	0.890 0.928 1.039	1996 2000 2004	1.188 1.216 0.863	1996 2000 2004	3.809 4.686 3.638
Total waterlogged area	1996 2000 2004	4.090 4.112 2.754	1996 2000 2004	2.052 1.654 2.043	1996 2000 2004	2.278 2.197 1.548	1996 2000 2004	8.393 7.963 6.344

Thus it may be concluded that there is a vast difference in the assessment of the extent of waterlogged areas by various agencies.

Waterlogged areas (post-monsoon) estimated by GWDUP & CGWB and RRSSC are summarized in Table 6.12 which suggests that the waterlogged area assessed by GDUP & CGWB is about 10 % higher than assessed by RRSSC, Jodhpur, though the reported area or assessed area by GWDUP & CGWB is about 3 million hectares less than the area assessed by RRSSC. So there may be underestimation by RRSSC or overestimation by GWDUP & CGWB of waterlogged areas based on groundwater level, which are 5.03560 million hectares and 6.344 million hectares, respectively. The average of these figures i.e., 5.689 million hectares can be taken as a reasonable waterlogged area in the state. It is 23.6 % of the geographical area of the state and 24.349 % of the command area of the state, which is quite high and calls for special attention to arrest the problem.

Table 6.12 Summary of waterlogged areas (post-monsoon) as estimated by GWDUP & CGWB and RRSSC

In million hectares

Category according to groundwater level	As assessed by RRSSC during 2003-05 (post-monsoon) within commands				As assessed by GWDUP & CGWB in 2004 (post-monsoon)		
	Geographical area	Command area	Waterlogged area	% of the command area	Reported area	Water logged area	% of the command area
1	2	3	4	5	6	7	8
	24.09	23.4007			20.23		
Most critical < 1 m			0.02959	0.126		} 2.706	13.337
Critical 1 to 2 m			0.9193	3.93			
Semi /less critical 2 to 3 m			4.0925	17.49		3.638	17.98
Subtotal			5.0360	21.52		6.344	31.36
Non-critical area > 3 m			18.359	78.45		13.88	68.64

Up till now, efforts to reclaim waterlogged areas had been limited to the construction of surface drains. However, if the water level is to be lowered, a combination of subsurface field drains and surface drains is an effective solution that needs to be tried in the state.

6.2 Crop Production

With an increase in irrigation facilities in the state, crop production also increased. The increase in total crop production and some of the major crops over the years is presented in Table 6.13 [16, 23].

Table 6.13 Increase in total food production and production of some of the major crops

Year	Total food production (million metric ton)	Food production of some of the major crops (million metric ton)			
		Rice	Wheat	Sugarcane	Potato
1	2	3	4	5	6
1984-85	29.913	7.153	15.675	70.888	5.449
2002-03	38.279	9.596	23.748	120.948	10.221
2011-12	52.057	13.917	32.150	126.11	11.997
2015-16	43.948	12.501	26.874	145.385	14.113
2016-17	55.747	14.467	34.971	156.949	13.956

Table 6.13 shows that between the years 1984-85 and 2016-17, total food production increased to the tune of 186 % in the state. Similarly, the production of major crops also increased substantially during this period.

During this period, the average productivity of major crops also increased manifold in the state as shown in Table 6.14 [16, 23].

Table 6.14 The average productivity of the major crops in the state

Name of the crop	The productivity of major crops in quintals per hectare					
	Years					
	1984-85	1997-98	1998-99	2011-12	2015-16	2016-17
1	2	3	4	5	6	7
Rice	13.00	21.42	19.36	23.58	21.33	24.15
Wheat	18.69	24.95	25.18	32.83	27.86	35.38
sugarcane	459.36	610.85	589.90	595.70	670.32	726.67
Potato	170.00	146.69	228.17	223.02	230.37	242.45

Table 6.14 shows that between the year 1984-85 and the year 2016-17 average yields of rice, wheat, sugarcane, and potato increased by 85.75 %, 89.29 %, 58.19 %, and 42.6 % respectively. Though, the yield of rice, wheat, and potato is still low in comparison to some of the leading states as shown in Table 6.15 [16, 23]. From table 6.15 it is clear that the average yield of rice,

wheat, and potato in the state is still low as compared to the average yield in Punjab and Haryana. But it can be said that the extension of irrigation facilities, particularly groundwater-based irrigation sources, which are considered an assured source of water supply by the farmers, encouraged them to invest more in agricultural inputs increasing crop productivity.

Table 6.15 Comparison of average productivity of major crops among leading states

S.no.	Name of state	Average productivity quintal per hectare					
		1997-98			2016-17		
		Rice	Wheat	Potato	Rice	Wheat	Potato
1	2	3	4	5	6	7	8
1.	Andhra Pradesh	24.31	-	-	35.40	-	-
2.	Haryana	28.00	36.60	218.91	32.13	45.14	249.12
3.	Punjab	34.65	38.53	155.76	39.98	47.07	256.96
4.	Tamilnadu	30.50	-	161.67	16.42	-	-
5.	Uttar Pradesh	21.48	24.95	146.69	24.15	35.38	242.45

7. WATER DEVELOPMENT FOR DOMESTIC/DRINKING SUPPLIES IN URBAN/RURAL AREAS

Urban water supply utilities in the state draw 1.992 BCM or about 2 BCM of groundwater annually. Apart from this, 653 major and small townships, spreading over 75 districts, are mainly dependent on groundwater. There are 5832 tube wells along with millions of India Mark II hand pumps, which are being used for providing drinking/domestic water supplies. Overall, 75 to 80 % of urban drinking water demand is dependent on groundwater resources [27].

8. POLICY INITIATIVES FOR WATER RESOURCES MANAGEMENT AND GOVERNANCE

Seeing the importance of Integrated Water Resources Management (IWRM) on account of an expected future water crisis, the state started policy initiatives as early as in the year 1999, and after that, several policy initiatives have been taken by the state government.

8.1 Policy Initiatives

8.1.1 State Water Policy (SWP)

State Water Policy adopted by the state government in the year 1999 was the first policy resolution of government making policy and strategy concerning water resources. Recognizing water resources as a state subject under the constitution, policy aligns with the general guidelines and parameters laid down in the   National Water Policy (NWP). It applies to all water resources of the state having the following objectives [2]:-

a.   Ensure the preservation of scarce water resources and optimize the utilization of available resources.

b.   Bring about qualitative improvement in water resource management, which should include the user's participation and decentralization of authority.

c.   Maintain water quality, both surface and underground, to established norms and standards.

d.   Promote the formulation of projects as far as and whenever possible on the concept of basins or sub-basins, treating both surface and the groundwater as a unitary resource, ensuring multipurpose use of the water resources.

e.   Ensure ecological and environmental balance while developing water resources.

f.   Promote equity and social justice among individuals and groups of users in resource allocation and management.

g.   Ensure self-sustainability in water resource development.

h.   Ensure Flood Management and drainage as an integral part of water resource development.

i.   Provide a substantive legal framework for management.

j.   Provide a Management Information System (MIS) for effective monitoring of policy implementation.

k.   Promote research and training facilities in the water resource sector.

l.   Provide a mechanism for the resolution of conflicts between various users.

SWP also consisted of an action plan, though various components of the action plan are yet to be implemented, showing the lack of commitment towards the management of water resources. For example, the creation of the State Water Planning Office (SWPO) was one of the crucial components of the action plan, but it is nowhere in the picture even after a lapse of more than 20 years. Similarly, other important provisions of the action plan, such as the enactment of legislation for regulation and control of surface and groundwater resources and their conjunctive use, administrative and legislative reforms for ensuring users participation in management and decentralization of authority, preparation of a perspective plan of the water resource development of the state on the integrated basis within the concept of basin/sub-basin development, etc. are yet to become a reality.

8.1.2 Observance of Ground Water Day and Ground Water Week

As stated earlier, irrigation and other sectors are heavily dependent on groundwater for their requirements, though, it is less than the surface water resource. Seeing the criticality of underground water in the development of the state, the government, in the year 2005 decided to observe Ground Water Day on the 10th day of June every year in the whole state to create awareness regarding the importance, judicious use, and management of groundwater. This day was observed in every Block and District as well as state headquarters. Later on, this was further strengthened by converting it into

Ground Water Week in the year 2012, and now Ground Water Week is observed in the whole state every year from 16th July to 22nd July. UP is the only state in the country that observes Ground Water Day/Ground Water Week every year. During the week, various awareness programs like poster, slogan writing, debate competition for students, rallies, workshops, seminars, exhibitions, etc. are organized with the active participation of all stakeholders [28].

8.1.3 Uttar Pradesh Participatory Irrigation Management Act, 2009

This act is promulgated to transfer management of the part of the canal system falling in the area of operation of water users' association. The act provides for delineation of area of operation of water users' association, the constitution of water users' association at the outlet, minor, distributary, and branch level of canal commands. Objectives of water users' association are [29]:-

(a) Promote and secure equitable, efficient, and timely water distribution.

(b) Motivate water users for adopting practices of scientific and economical use of water.

(c) Encourage conjunctive use of surface and groundwater.

(d) Encourage intensified and diversified agricultural production systems.

(e) Protect the environment and ecology.

Along with other provisions, the Act also provides for the election of the representatives of sub-commands and chairperson and office bearers of managing committee at outlet, minor, distributary and branch level, transfer of management of irrigation system to water users' association, penal provision of imprisonment up to six months or fine of minimum INR 1000 extended up to the cost of damage or both.

However, the act neither specifies the tenure of elected members leaving it to decide later on by competent authority nor any time limit for the notification of operationalization of water users' association. So the act lacks to ensure timely constitution and regular election, which dilutes the efficacy of the act. Sluggish progress of the constitution of the water users' association supports this view. After ten years of the inception of the act, water users' association on all the outlets, minor, distributary, and branch level should have been constituted up till now.

8.1.4 Policy for Groundwater Management, Rainwater Harvesting, and Groundwater Recharge

The State Government came out with a comprehensive policy document in February 2013 for overall groundwater management. This document named "Policy for Ground Water Management, Rain Water Harvesting and Ground Water Recharge in the State of Uttar Pradesh" covers almost all important aspects of groundwater resource management, including its regulated development/extraction, optimum use, and conservation as well as detailed mapping of aquifers. The priority areas emphasized in the policy for further action are [30]:-

i) Aquifer mapping and aquifer-based groundwater management.

ii) Optimum use of groundwater and planned exploitation.

iii) Integrated rainwater harvesting and groundwater recharge.

iv) Setting the groundwater regulation process.

v) Monitoring and mapping of groundwater quality for environment protection.

vi) Promoting groundwater data management by establishing State Groundwater Informatics Centre.

vii) Preparation of district-wise water management plans.

viii) Training, publicity, extension, and public awareness.

ix) Strengthening the existing institutional system.

UP is the first state to have such a policy in India. Though it is a very comprehensive & diversified policy document, the implementation of its various provisions had been slow, and it could not bring much difference in overall groundwater management in the state.

8.1.5 The Uttar Pradesh Water Management and Regulatory Commission Act, 2014

This act has been promulgated for the establishment of Uttar Pradesh Water Management and Regulatory Commission to regulate and recommend the tariff for water used for agriculture, industrial, drinking, power, and other purposes and also for levying cess on land benefitted by flood protection and drainage works from the owners of the land. The powers and functions of the commission are following [31]:-

(a) to fix and regulate the water tariff system and   charges for surface and sub-surface water used for domestic, agriculture, industrial, and other purposes;

(b) to determine and regulate the distribution of entitlement for various categories of users as well as within every category of use;

(c) to periodically review and monitor the water sector costs and revenues;

(d) to determine the rate of cess to be charged from the owner of lands benefitted by flood protection and drainage works implemented under new projects;

(e) to advise the State Government for the Integrated State Water Plan/Basin Plans developed by State Water Resource Agency to ensure sustainable management of water resources within the parameters laid down by State Water Policy as amended from time to time;

(f) to determine the allocation and distribution of entitlements for various sectoral water requirements as per the State Water Policy;

(g) to review and accord clearance to new water resource projects proposed at the river basin / sub-basin level by the concerned entity ensuring that the proposal conforms with the Integrated State Water Plan, especially, concerning water allocation of each entity, that is economical, hydro-geologically and environmentally viable;

(h)   to establish a system of enforcement, monitoring, and measurement of the entitlement for the use of water to ensure that the actual use of water complies with the entitlements;

(i)   to monitor the conservation of the environment and facilitate the development of a framework for the preservation and protection of the quality of surface and groundwater resources as per the established norms and standards;

(j)   to promote competition, efficiency, and economy in the activities of the water and wastewater sector to minimize wastage of water;

(k)   to promote better water management techniques and practices;

(l)   to enforce rainwater harvesting to augment groundwater recharge;

(m)   to enforce the decision or orders issued under this act by a suitable agency for the purpose;

(n)   to aid and advise the State Government on any matter referred to the Commission by the State Government;

The above powers and functions from (d) to (n) are recommendatory/ advisory and cannot take place without the approval of the competent authority of the State Government.

The above act was promulgated initially in the year 2008 but abolished later on. It was revised and re-enacted in the year 2014. Its implementation had been sluggish as the Chairman of the Commission, constituted under the act, could be appointed in the year 2018, and its other members and support staffs are yet to be appointed. Rules to operationalize the provisions of the act are also yet to be framed. In a nutshell, it can be said that the Commission has not been able to discharge any of its functions. Most of the powers of the Commission are advisory, and there is no clause for offenses and penalties, making it less effective.

8.1.6 The Uttar Pradesh Ground Water (Management and Regulation) Act, 2019

The Government of Uttar Pradesh recently enacted "The Uttar Pradesh Ground Water (Management and Regulation) Act, 2019" to provide for protecting, conserving, controlling, and regulating groundwater to ensure its sustainable management in the State, both quantitatively and qualitatively, especially in stressed rural and urban areas.

The State Government has brought this new legal framework in the public interest to manage and regulate the extraction and use of groundwater judiciously in any form and also to conserve and protect groundwater in the stressed areas of the state. The rules under this act have also been framed as the Uttar Pradesh Ground Water (Management and Regulation) Rules, 2020, and are in the process of operationalizing. Key provisions of the act are the following [32]:-

• Powers to notify areas for management and regulation of groundwater resources are vested in the State Government.

• Act provides to regulate industrial, commercial, infrastructural, and bulk users of groundwater only, and they are required to apply for a grant of certificate of registration.

• Construction of new wells for commercial, industrial, infrastructure, and bulk use, including the construction of borings/tube wells under government schemes except for government schemes for drinking water supplies and tree plantations is banned in notified areas.

• Penal provision shall not apply to agriculture and domestic users of groundwater, which form the majority of groundwater users; however, they are required to register online or directly to the respective Block Panchayat Ground Water Management Committee/Municipal Water Management Committee for groundwater usages. They are also required to self-regulate.

• To carry out the functions under the act, the following Institutional Framework is provided in the act-

(a) Gram Panchayat Ground Water Sub-Committee in every Gram Panchayat (Village level elected body) shall be the lowest unit in the rural areas under the chairmanship of Gram Pradhan (village head). It shall prepare the Gram Panchayat Groundwater Security Plan.

(b) Block Panchayat (Block level elected body) Ground Water Management Committee shall work at block level under the chairmanship of Block Pramukh (Blockhead) and prepare Block Level Ground Water Security Plan by consolidating gram panchayat level plans. The committee shall also monitor the implementation of the Groundwater Security Plan. The Block Committee shall register all the wells except those of existing commercial, industrial, infrastructural and bulk users.

(c) Municipal Water Management Committee shall be the lowest public unit for managing groundwater in urban areas under the chairmanship of Nagar Pramukh/Nagar Palika Pramukh (Municipal head). The committee shall prepare an overall Municipal Groundwater Security Plan and monitor its implementation. The committee shall register all wells other than those of existing commercial, industrial, infrastructural, and bulk users.

(d) District Groundwater Management Council shall be an overall unit for the management of groundwater resources at the district level under the chairmanship of the District Magistrate. The District Council shall consolidate Block Panchayat and Municipal Groundwater Security Plans into district-level Groundwater Security Plan and ensure its implementation. The council has powers to grant a certificate of registration to existing commercial, industrial, infrastructural and bulk users of groundwater in notified areas and existing and future commercial, industrial, infrastructural and bulk users of groundwater in non-notified areas. It also has the powers to grant no-objection certificates for the extraction of groundwater to these users.

(e) The State Groundwater Management and Regulatory Authority shall be the apex institution under the chairmanship of the Chief Secretary, Government of UP. The State Authority shall notify the areas for groundwater management and regulation, fix groundwater abstraction limits, and take groundwater pollution control measures.

- In rural areas, over-exploited and critical blocks, whereas stressed urban areas with a significant decline of groundwater levels, i.e., more than 20 cm/year recorded during the last five years are to be designated as Notified Areas for the regulation.

- Groundwater abstraction limits are to be fixed for the notified and non-notified areas, based on the hydrogeological conditions and resource potential of the area concerned.

- Fee on groundwater extraction shall be charged annually from commercial, industrial, infrastructural, and bulk users of groundwater based on the quantity of groundwater extracted.

- For prevention of groundwater pollution following measures are provided-

  - Groundwater Quality Sensitive Zones shall be declared.

  - Prohibition on disposal of polluting matter into the well.

  - Ban on direct recharging through the water collected from open areas, except for rooftops into the aquifers.

  - Prohibition on polluting ponds, rivers, wells.

- Provision of self-regulation by the farmers in notified areas. Rainwater harvesting shall be an integral part of water security plans. Rooftop rainwater harvesting and combined recharge system in municipal areas shall be enforced as per provisions of existing building by-laws. Recycling and reuse of groundwater shall be encouraged.

- Impact assessment and transparency shall be ensured.

- Penalties up to the imprisonment of seven years and a fine up to INR 2 million.

- Provision to create Groundwater Fund to be utilized for groundwater management activities.

8.1.7 Incorporation of a chapter on rainwater harvesting in School Syllabus

To sensitize and educate the students on the importance of water resources and their conservation, the Government of UP introduced rainwater harvesting as a chapter in the syllabus of 6th to 12th class. The chapter broadly includes the concept, significance, and benefits of rainwater harvesting and groundwater recharge in the overall conservation and management of water resources. The suitable techniques and the appropriate methods of rainwater harvesting and groundwater recharge are also discussed in the chapter.

8.1.8 Executive orders regarding Rainwater Harvesting and Groundwater Recharge

In UP, the State Government has always been proactive and felt the need for "Rain Water Harvesting & Ground Water Recharge" (RWH & GWR) in the stressed areas of the state to rejuvenate the depleting aquifers and improve groundwater availability. Various initiatives & policy decisions, envisaging suitable provisions for rainwater harvesting, have been taken at the government level and elaborate government orders have been issued from time to time with detailed technical guidelines and executive provisions for compliance and execution. Some critical clauses made through executive orders are narrated below [28]:-

• Provisions were made for rainwater harvesting and groundwater recharge in urban areas way back in 2001, which were amended from time to time. Later on, these provisions were also included in the Building Construction and Development By-Laws, 2000.

• Executive Committee, under the chairmanship of the Chief Secretary, UP, was constituted in the year 2004 to review the rainwater harvesting schemes in the state.

- Ground Water Department UP was declared as "Nodal Agency" for groundwater management and monitoring of rainwater harvesting schemes in the year 2004.

- Technical Co-ordination Committee (TCC) under the chairmanship of District Magistrate was constituted in all the districts for implementation and monitoring of Rain Water Harvesting Projects/Schemes.

- Rainwater harvesting was made mandatory for all new housing & group housing schemes in the urban areas of the state.

- Rooftop rainwater harvesting is compulsory for a plot size of 300 square meters or more. The provision has also been included in the Building By-Laws of the Housing and Urban Planning Department.

- For all the government buildings (both new & old), the installation of the Rooftop Rainwater Harvesting System is mandatory since the year 2001.

- Provision of a combined groundwater recharge system is now mandatory for all group housing schemes and the plots below 300 square meters, in all the new housing schemes as the collective/combined system of recharge is the most suitable, easy, and cost-effective method for rooftop rainwater harvesting and groundwater recharge in urban areas. This system envisages the installation of a separate rainwater pipeline for all the group houses in the new housing schemes, which is joined to a common recharge/storage facility for rainwater harvesting and groundwater recharge.

- Direct recharging of rainwater to aquifers from open/paved/unpaved areas was prohibited due to the risk of pollution/contamination which is now included in The Uttar Pradesh Ground Water (Management and Regulation) Act, 2019 with strict penal provisions.

- Conservation of existing water bodies in all urban areas was made mandatory.

- The allocation of 5 % of land for the development of water bodies was made mandatory in all the new housing schemes.

- Identification of natural catchment & feasibility assessment was made an integral part of pond development/rejuvenation.

- Recharge shaft was prohibited in ponds as the risk of industrial/other pollution may occur.

- In parks, only 5 % area is allowed to be covered with concrete/ pavements.

- It was made mandatory that roadside footpaths should be covered as far as possible with perforated bricks on edge/loose stone tiles to allow maximum natural recharge of groundwater.

- In shallow water level areas, rainwater from roof catchments should be conserved as surface storage.

- Area-specific hydrogeological parameters and feasibility assessment should be adopted for the design and implementation of rainwater harvesting structures.

- Maintenance of the recharge structures should be an integral part of all the recharge schemes.

- Impact Assessment should be ensured according to the guidelines issued by the Nodal Agency (Ground Water Department, UP) in the year 2004.

8.1.9 Ban on Irrigation Tube wells

A complete ban on all kinds of tube wells/bore wells constructed for irrigation purposes under all government schemes was imposed to control groundwater exploitation in over-exploited and critical blocks. These include all tube wells constructed by government departments and tube wells constructed by individuals availing government subsidies [28].

8.1.10 Bhujal Sena (Groundwater Army)

The concept of Bhujal Sena was introduced in the year 2017 in all the districts of the state. It is an independent and dedicated group of local people, with the sole aim to create and launch a public awareness campaign continuously for groundwater management and conservation in the entire state. Representatives of NGOs, students, teachers, environmentalists, social workers, and scientists are included in this group. Fifty such members are nominated in each district level Bhujal Sena by the respective District Magistrates. Bhujal Sena aims to work as per the Action Plan prepared annually by the Ground Water Department, UP to be monitored through District Magistrates [33].

In principle, Bhujal Sena seems a very innovative initiative of the State Government. With a total strength of 3750 Bhujal Sainiks (Groundwater soldiers) in the state, the Bhujal Sena could be successfully utilized for creating effective awareness [33]. But due to a lack of infrastructure and motivation for implementing government agencies, this program has not taken off as planned at the time of inception.

8.2 Initiatives through new schemes

After the adoption of SWP, the state started several initiatives through new schemes to implement the water management agenda. Some important new schemes are discussed in the following sections.

8.2.1 Uttar Pradesh Water Sector Restructuring Project (UPWSRP)

Assisted by the World Bank, this project was launched in the year 2002 to introduce water sector reforms through a programmatic framework. The project was proposed to implement the provisions of SWP.

First Phase of UPWSRP

The first phase of the project was started in March 2002 with a prime expenditure (the year 2002) of INR 8190.31 million. Under the Irrigation and Drainage Sector Development Program of the first phase, rehabilitation and modernization of canal system and drains were carried out in o.343 million hectares of command area including renovation of the part of Ghagra-Gomti Haidergarh Branch canal and Jaunpur Branch. Rehabilitation and modernization work in the canal include clay work in 2556 km of the canal, 39 duckbill weir, 5 cross regulators, 463 gates, 110 km parallel minor, 2710 other concrete works, and 2369 km clay work, and 398 concrete works in drains.

Under Water Sector Reforms, State Water Resource Agency (SWaRA), State Water Resource Data and Analysis Centre (SWaRDAC), Uttar Pradesh Water Management, and Regulatory Commission (UPWaMReC), and Jaunpur

Branch Sub-basin Development and Management Board were constituted. Uttar Pradesh Participatory Irrigation Management Act, 2009 was also promulgated in 2009 followed by notification of rules in 2010 and registration and election process of 8880 outlet level, 407 minor level, 398 federated minor level, and 28 distributary level water users' association. Data Centre was also established in the Irrigation Department.

Five hundred piezometers fitted with an automatic water level recorder and 384 piezometers for manual recording were installed in the project area to monitor groundwater levels. Apart from this, 91 automatic level machines were arranged and 26 online Automatic Weather Stations were installed, 9 lakes were renovated, 18 check dams were constructed, 18 lakes were developed for fisheries, 1949 Women Self Help Groups were formed, and demonstrations and training for modernization and diversification of agriculture were conducted [7].

World Bank assessed the achievements of the first phase as moderate. Probably there is no independent third-party evaluation of the achievements of the first phase. Most of the automatic water level recorders became defunct after some time, and no information is available on how the obtained data from these were analyzed and used in decision making. Likewise, the impact of demonstrations and training for agriculture diversification, the constitution of water users association, the impact of the creation of a new organization, etc. are not known.

Second Phase of UPWSRP

The second phase of the project started in October 2013 and its end date is October 2020. The Project Development Objectives are; i) strengthen the institutional and policy framework for integrated water resources management for the entire State; and (ii) enable farmers in targeted irrigated areas to increase their agricultural productivity and water use efficiency.

The project area is tail from 22.98 km of Lower Ganga Canal System, Haidergarh Branch and Rohani, Jamini and Sajnam Dam Canal system covering sixteen districts i.e. Barabanki, Raebareli, Amethi, Lalitpur, Etah, Firozabad,

Kasganj, Mainpuri, Farrukhabad, Etawah, Kannauj, Auraiya, Kanpur Dehat, Kanpur Nagar, Fatehpur, and Kaushambi. The World Bank has approved the total fund of INR 28350.00 million. Works of the second phase are rehabilitation and modernization of selected canal system, the constitution of water users' association, preparation of basin plans, modernization of Naraura Barrage, strengthen groundwater monitoring network, capacity building of groundwater department, demonstration of aquifer management and conjunctive use through pilots, etc. [7].

8.2.2 Redevelopment and Management of Large Ponds

Plains forming the major portion of the state are gentle sloppy due to which ponds had been the traditional and effective rainwater harvesting, water storage, and groundwater recharge structures in the state. Traditionally, every village had more than one pond, and they are still plenty in the state. But after independence due to preoccupation to develop water to fulfill the growing demand for irrigation and other sectors, these ponds were overlooked, and no attention was paid to their maintenance and upkeep. In ancient times, these were also the religious-cultural centers of the community, and the community use to preserve these as water sanctuaries. The neglect of these ponds ultimately resulted in reducing their capacity to harvest and store rainwater substantially due to heavy siltation over time. Many of these were encroached upon due to urbanization and the expansion of agricultural activities and exist only on paper. The major thrust for their renovation and de-siltation came when rainwater harvesting works were given priority in the centrally sponsored Mahatma Gandhi Rural Employment Guarantee Scheme, which was launched by the Government of India in the year 2006. Large numbers of small-size ponds were de-silted under this scheme in the state. But, being an employment guarantee scheme, the use of machinery is forbidden in it due to which, large-size ponds (> 01 hectare) requiring heavy earth movement could not be de-silted under it.

Though authentic and exact information about the numbers of ponds in the state is not available, according to an estimate, about 0.025 million ponds are above one hectare, having an average area of about 2.5 hectares. Desilting these ponds up to 3 m or more requires heavy earth movement, and the cost of redeveloping these ponds comes about INR 1.0 to 1.2 million per hectare. To redevelop these ponds, State Government started a new scheme "Redevelopment and Management of Ponds" in the year 2018-19, so that their water storage and recharge potential is restored and they are redeveloped as water sanctuaries.

One innovation in this scheme is that the concept of "Pani Panchayat" to ensure public cooperation is introduced in it right from planning to execution. Pani Panchayat is a group of village women and men who are involved from the beginning with the following objectives [34]:-

i)   increase community participation for maintenance of water bodies and the creation of funds at the village level;

ii)   make efforts to increase the rainwater storage and increase the water storage capacity of traditional water bodies;

iii)   protection of redeveloped ponds and increase means of livelihood through the plantation, fisheries, and pottery;

iv)   ensure the first right of the woman on water through safe equity, local institutional development, and capacity building;

v)   initiate community water management through decentralization;

vi)   create community awareness about the over-abstraction of groundwater and ensure judicious use of water by promoting low water consuming crops;

vii)   promote women's leadership by involving women advocacy groups in the process of decision-making of Panchayati Raj Institutions regarding the use and share of water for agriculture purposes.

Initial results were encouraging, and a sort of movement started to build around these ponds to save and manage water. Even it acted as third-party concurrent feedback giving information about the quality of work, possible embezzlements, etc. These groups also helped in removing the encroachment and resolving disputes. But these groups have no legal right on the pond and its water hindering their sustainability. The allocation of funds is also not sufficient in the scheme according to the requirement as yearly allocation is only INR 480 million, which is sufficient to develop about 430 numbers of ponds only. With this allocation, it will take about 60 years to redevelop all the ponds.

This scheme can make a difference in the future if sufficient funds are allocated for it, and "Pani Panchayat" is recognized as a legal entity having right over the pond and its water.

8.2.3 National Hydrology Project

To develop a hydrological monitoring network in the country, the Government of India implemented Hydrology Project– I (the year 1995 to the year 2003) and Hydrology Project –II (the year 2006 to the year 2014) with the assistance of the World Bank. Uttar Pradesh was not a part of these projects. The Government of India further extended these efforts into the National Hydrology Project (NHP) covering the whole country for which a 50% grant will be provided by the World Bank and 50 % by the Government of India. The objective is to develop a state-of-the-art Hydrological Information System in the country. Concepts of the NHP are [7]:-

(1) Standardization of water resource monitoring and information system for the country with uniform procedures and databases.

(2) Enhancing data exchange between center and states.

(3) Improving access to information in the public domain.

(4) Introducing countrywide generic solutions for flood forecasting and water resource management.

(5) Developing site specific-solutions for water resource planning, operation, and management, including remote sensing-based techniques.

Major activities proposed under the project are the following:-

(a) Expanding and upgrading of water resource monitoring and data acquisition system (including real-time system).

(b) Developing and supporting a centralized database management system, water resource data set including remote sensing-based information, and facilitate State Water Resources Information Systems.

(c) Developing Decision Support System (DSS) in selected river basins for flood forecasting and reservoir operation, water resource (Surface water and Groundwater) planning and management, etc.

(d) Capacity building through establishing and supporting water resources data centers, training, centers of excellence.

In the state of U P, proposed works under NHP (Surface Water) costing INR 10500.00 million are; I) 24 Supervisory Control and Data Acquisition (SCADA) on Barrages including Regulators, ii) 18 SCADA on Dams, iii) 94 Hydrological Information Systems on Dams, iv) 551 Discharge & Rain Gauge (DRG)/Automatic Rain Gauge (ARG) stations in various blocks of the state, v) 119 Automatic Weather Stations at dams and vi) 71 Gauge Discharge Stations.

Under NHP (Ground Water) also, Ground Water Resource Data Acquisition System, Ground Water Resource Information System, Ground Water Resource Planning and Operation, and Institutional Development and Capacity Building are proposed. For continuous data acquisition, automatic groundwater level recorders or digital water level recorders are being installed. The focus of activities under this component is in Bundelkhand and Western regions of the state [35].

8.2.4 Atal Bhujal Yojana

Recently (2019) Jal Shakti Ministry (Ministry of Water Resources), the Government of India launched an ambitious National groundwater management improvement scheme named as "Atal Bhujal Scheme" under Jal Jeevan Mission. The objective of the scheme is to efficiently improve groundwater conditions through area-specific interventions and community participation in the problematic areas of the country. The focus of the schemes will be to emphasize the recharge of groundwater resources and the use of water-efficient methods to reduce groundwater exploitation.

The scheme will be funded by the Government of India and the World Bank on fifty-fifty percent sharing. It will be implemented in 7 states; Uttar Pradesh, Madhya Pradesh, Rajasthan, Karnataka, Maharastra, Gujrat, and Haryana across 78 districts, 193 blocks, and about 8300 Gram Panchayats (Village level units). In U P, 26 blocks, 20 from the Bundelkhand region, and 6 from western UP are selected. The scheme is a performance-based program. Water security plans will be implemented at the gram panchayat level, and the success of the scheme will be monitored by observing improvement in groundwater conditions. This is an incentive-linked program for the community to achieve groundwater management targets [36]. Seeing the large numbers of over-exploited and critical blocks coverage of the scheme seems too less.

8.3 Organizational Change

In the state, water establishments/departments were created long back when the development of water was the main goal to increase irrigation facilities and, in turn, crop production. These are mainly manned by engineers. The water sector now faces new challenges and required a multidisciplinary team to meet these challenges. With the declaration of SWP, restructuring of water establishments also started in the state, though at a very slow pace.

Water resources are being planned, developed, and managed by different departments in the state. Surface water, including the canal and river, is managed by the Irrigation department including major and medium-lift schemes. Beyond outlet, canal water is mainly managed by farmers with little intervention of Command Area Development in some areas. Estimation and exploration of groundwater resources are carried out by the Ground Water Department, whereas the development of the resource is being done by the Irrigation Department, Minor Irrigation Department, and UP Jal Nigam.

The utilization of water is controlled by various departments viz. Agriculture, Urban Development, Rural Development, Power, Industries, Tourism, Environment, Pollution Control, Forests, and others [37].

To manage the water resources in an integrated manner, State Government, constituted State Water Board (SWB) under the Chairmanship of Chief Secretary, Government of UP to prepare policy and program, establish coordination between various departments/organizations dealing with water management for optimal use of water resources. SWB consisted of Additional Chief Secretary (Uttarakhand), Agriculture Production Commissioner, UP, Principal Secretary Irrigation, UP, Principal Secretary Energy, UP, Principal Secretary Finance, UP, Principal Secretary Planning, UP, Principal Secretary Industry, UP, Chief Engineer Irrigation, UP, Managing Director, Jal Nigam, UP, Director, Ground Water, UP, Nominated member of Central Water Commission, Nominated members of Central Ground Water Board as members and Chief Engineer (Design & Planning), UP as Member Secretary [37]. However, SWB remained more or less non-functional.

To assist the State Water Board and strengthen Socially & Environmentally sustainable multi-sectoral Water Resource Planning Allocation & Management Capacity, the following apex level water institutions were created in Uttar Pradesh under the water sector reform component of the World Bank-funded UP Water Sector Re-Structuring Project [37]:-

(a) UP State Water Resources Agency (SWaRA)

(b) UP State Water Resources Data Analysis Centre (SWaRDAC)

(c) UP Water Management and Regulatory Commission

8.3.1 Establishment of State Water Resources Agency (SWaRA)

State Water Resources Agency (SWaRA) was created for management, planning, and sectoral allocation of water resources (both surface and underground) to various agencies viz. Drinking-Water, Agriculture, Industrial Development, Hydro Power, Transportation, Entertainment, Thermal Power Production, and Environmental flow of water in the rivers, and to give a legal base to the above purposes and use of surface water/groundwater as well as to work as technical secretariat to the State Water Board.

However, it could not be made adequately functional since its creation in the year 2001 and always struggled with the problem of human resources. It is evident from the fact that after the lapse of about 20 years, it could prepare the Integrated Basin Plan of only 5 major basins in the state up to March 2020. Even then, these plans are theoretical and further need activity-wise and department-wise micro-planning with necessary financial commitments to make them implementable.

8.3.2 Establishment of State Water Resource Data Analyses Center

The objective of establishing the State Water Resources Data Analyses Centre (SWaRDAC) was to provide water-related data to the SWaRA and other central/state water plans for all river basins of the state. However, it could not create an effective mechanism to retrieve data from line departments and its regular up-gradation. It also faced a shortage of manpower in general and had very sluggish working since its inception. Ultimately, after about 20 years of its creation, it could compile basin-wise data for the preparation of basin plans.

8.3.3 Establishment of Water Management and Regulatory Commission

Initially, Water Management and Regulatory Commission were established under The Uttar Pradesh Water Management and Regulatory Commission Act, 2008. Later on, this act was abolished, and after modification, it was re-enacted in the year 2014 as "The Uttar Pradesh Water Management and Regulatory Commission Act, 2104" and Commission was reconstituted in the year 2014 [38]. However, under the provisions of this act, the Chairman of the Water and Regulatory Commission could be appointed only in the year 2018 [39]. Other members of the Commission could not be appointed up to the end of the financial year 2019-20. Commission also faced a shortage of manpower regularly. Some of the main functions of the

Commission were to regulate water tariff systems, charges for surface water and groundwater used for domestic, agriculture, industrial, and other purposes, determine and decide entitlement of water for various usages, review and monitor the costs of the water sector and revenue, etc. but it could not discharge any of these functions and has been more or less a nonfunctional entity.

8.3.4 Reorganization of Water Establishments and Ministry

In its efforts to further integrate the water governance in the state, the Government of UP integrated water establishments at the apex level by bringing the Minor Irrigation Department and the Ground Water Department under Namami Gange and Rural Water Supply Department in the financial year 2019-20 [40]. Jalshakti Mantralaya (Water Power Ministry) was also consolidated by bringing this new unified department under it. Now, most of the departments engaged in water development and management i.e. Department of Irrigation & Water Resources, Minor Irrigation, Ground Water, Namami Gangey & Rural Water Supply are under one Ministry. But nothing has changed down the line.

9. CHALLENGES

With ever-increasing water demand due to the growing population and looming climate change complications, the water sector faces new challenges at present, and different strategies and out of box solutions are required to deal with the situation. The following sections describe the challenges the water sector faces at present in the state.

9.1 Diminishing Rainfall

The average annual rainfall is continuously diminishing in UP. The diminishing rate of the average annual rainfall is, however, faster in the Bundelkhand region than in the state as a whole, as may be seen from Table 9.1 and figure 3. In the state of UP, the average annual rainfall during the 2001-2010 decade was 17.77 percent less than the long-term average annual rainfall. In the Bundelkhand region, the shortfall was 24.44 percent, which means that the decline in rainfall in the Bundelkhand region was more pronounced than the state as a whole. This also implies that drought-like conditions prevailed in this region of state during this decade. There are several manifestations of drought-like conditions including late arrival of rains, early withdrawal, a long break in between, insufficient water in reservoirs, and drying up of wells leading to crop failures and even un-sowing of the crops which ultimately hampers livelihood[41].

Table 9.1 Decennial Mean Annual Rainfall in Uttar Pradesh, and Bundelkhand region of the State of Uttar Pradesh, India

S.no.	Decade	Average annual rainfall (mm)		Long term mean annual rainfall of the state (mm)
		State	Bundelkhand Region	
1	2	3	4	5
1.	1981-1990	911.77	799.2	
2.	1991-2000	870.27	781.3	
3.	2001-2010	801.74	709.2	
				974.4

Data Source: State Water Resources Agency, Uttar Pradesh, India

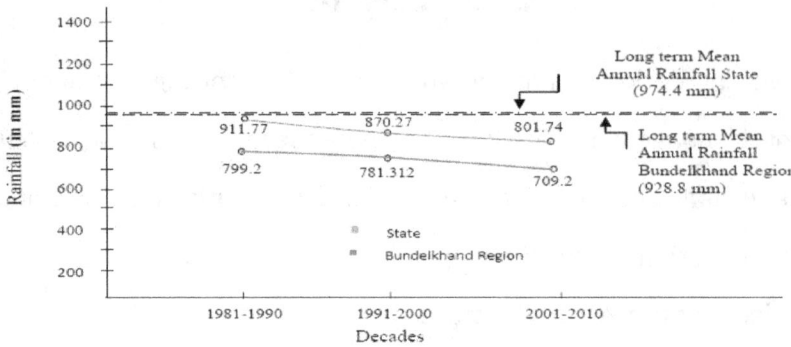

Fig. 3   Decennial Mean Annual Rainfall in Uttar Pradesh, and Bundelkhand

The frequency of drought is also increasing. An analysis of the Bundelkhand region, which is a parched region of the state and known for its water woes due to its peculiarities such as rocky underground formations, high slops, deforestation, etc. shows that during the 18th and the 19th Centuries, the Bundelkhand region had a draught, on average at every 16 years which increased three times during the period 1968-1992. The most recent and the continued period of poor rainfall recorded in the region was during 2004-2010 when below average and erratic rains were recorded in most parts of the region [42, 43]. However, the region received above-normal rainfall in 2011 but, after 2011, the rainfall again turned deficient and erratic. Analysis of rainfall data for the seven districts of the region shows that, in almost all districts of the region, there was a negative deviation in the rainfall during the period 2013-2017 (Table 9.2). It is clear from Table 9.2 that the year 2013 was a good rainfall year in which all the districts of the region received above-normal rainfall. Years 2014 and 2015, however, were bad rainfall years in which all the districts of the region received below normal rainfall. The situation was somewhat better in the year 2016 as three districts received above-average rainfall, but the situation deteriorated again in the year 2017 as all the districts received below-average rainfall. The deficiency in the average annual rainfall was particularly severe in the Mahoba district. During the year 2015, the deficiency in the average annual rainfall in the district was almost 80 percent, which had a very serious and long-lasting impact on the livelihood opportunities in the district. In other districts of the region also, the shortfall in the average annual rainfall was quite substantial during the period.

Table 9.2 Shortfall in the average annual rainfall in Bundelkhand region

District	Rainfall)/ Deviation (mm/%)	2013	2014	2015	2016	2017	5 years Average
1	2	3	4	5	6	7	8
Lalitpur	Actual	1684.7	810.9	449.2	919.4	674	907.64
	Average	1034.6	1034.6	1034.6	1034.6	1034.6	1034.6
	Deviation	62.8	- 21.62	-56.6	- 11.1	- 34.8	- 12.27
Jhansi	Actual	1204.1	569	533.5	640.4	468.7	683.14
	Average	931.5	931.5	931.5	931.5	931.5	931.5
	Deviation	29.26	- 38.9	-42.7	- 31.2	- 49.6	- 26.66
Jalaun	Actual	894.8	502.7	450.1	700.14	361.8	581.96
	Average	871.5	871.5	871.5	871.5	871.5	871.5
	Deviation	2.67	-42.31	-48.3	19.63	58.48	- 33.22
Mahoba	Actual	868.8	379.8	181.2	843.3	274.3	509.48
	Average	853.1	853.1	853.1	853.1	853.1	853.1
	Deviation	1.84	-59.48	-78.7	-1.1	-67.8	-40.27
Hamirpur	Actual	1136.5	485.1	470.4	960.6	380	686.52
	Average	881.8	881.8	881.8	881.8	881.8	881.8
	Deviation	28.88	-44.98	-46.7	+8.9	-56.9	-22.14
Banda	Actual	984.6	639.2	676.9	1230.5	508.3	867.9
	Average	943.2	943.2	943.2	943.2	943.2	943.2
	Deviation	4.389	-32.33	-28.2	+30.4	- 46.1	-14.34
Chitrakoot	Actual	1318.3	665.9	531.1	1073	834.5	884.56
	Average	986.3	986.3	986.3	986.3	986.3	986.3
	Deviation	33.66	-32.48	-46.1	+8.79	-15.39	-10.3

Data source: Customized Rainfall Information System [43]

From the above analysis, it can be concluded that rainfall is no more a reliable source of water supplies, and it is expected that looming climate change complications will disturb the balance of the present distribution of water, making rainfall more unreliable. So to mitigate the effects of climate change and to meet the future requirements of a growing population, search for new sources of water supply becomes imperative.

Wastewater/sewage water may be such a source as it is available round the year in the same quantity.

9.2 Inadequate Forest Cover

The Forest Policy of the nation declared after independence envisaged to develop forest in one-third or 33% of the land area. But in the state of Uttar Pradesh, India, and its Bundelkhand region, forest cover remained far below this target. After the formation of the new state of Uttarakhand, separating the hilly region of UP in the year 2000, forest cover came down drastically and remained stagnant around 6% in the state and around 6.5 % in the parched Bundelkhand region, far below the targeted one-third of the area. Forest cover in the state of UP, since 2001 is shown in Table 9.3.

Table 9.3 Forest cover in Uttar Pradesh, India

S. no.	Assessment Year	Forest cover in % of Geographical Area	
		State	Bundelkhand Region
1	2001	5.71	6.27
2	2003	5.86	6.62
3	2005	5.86	6.62
4	2009	5.95	6.66
5	2011	5.95	6.66
6	2013	5.96	6.77
7	2015	6.00	6.72
8	2017	6.09	6.55

Data source: State of the forest Report [44]

The relation between vegetation and precipitations (rainfall/snowfall) had been a debatable issue. But in the recent past, a consensus is again developing that there is a relation between precipitation and vegetation. Global Landscapes Forum conducted a session on "Precipitation and its relation to vegetation" to show the role of forest and trees in the water cycle and how to influence the climate through atmospheric water cycle controls

[45]. A new hypothesis suggests that forest cover plays a much more significant role in determining rainfall than previously recognized. It explains how forested regions generate large-scale flows in atmospheric water vapors. Under this hypothesis, high rainfall occurs in continental interiors such as Amazon and Congo River basins only because of near-continuous forest cover from interior cost. The underlying mechanism emphasizes the role of evaporation and condensation in generating atmospheric pressure differences and accounts for several phenomena neglected by existing models. It suggests that even localized forest loss can sometimes flip a wet continent to arid conditions. If it survives scrutiny this hypothesis will transform how we view forest loss, climate change, hydrology, and environmental services [46].

To scrutinize the above hypothesis in the Bundelkhand region of the state, the relation between five years' average deviation in rainfall during the year 2013 - 2017 and forest cover (2017 assessment) of seven districts of Bundelkhand Region is shown in Table 9.4 and figure 4.

Table 9.4 Forest cover and deviation in rainfall

S.no.	Name of district	Forest cover (%) 2107 Assessment	Rainfall deviation (%) 5 years average
1	2	3	4
1.	Lalitpur	11.35	-12.27
2.	Jhansi	6.03	-26.66
3.	Jalaun	5.45	-33.22
4.	Mahoba&Hamirpur	5.54	-31.00
5.	Banda&Chitrkoot	9.01	-9.17

Note: Moahoba and Chitrakoot Districts were formed by bifurcating Hamirpur and Banda districts respectively

Table 9.4 indicates a clear correlation between forest cover and deviation in rainfall, though this relationship is not linear but supports the above hypothesis as rainfall deviations are more where forest cover is less and vice versa. It can also be concluded that in these districts, had the forest cover been of one-third of the area, the effects of climate change on rainfall would have minimized. Seeing the previous record, increasing forest cover is an uphill task and is one of the biggest challenges for the state.

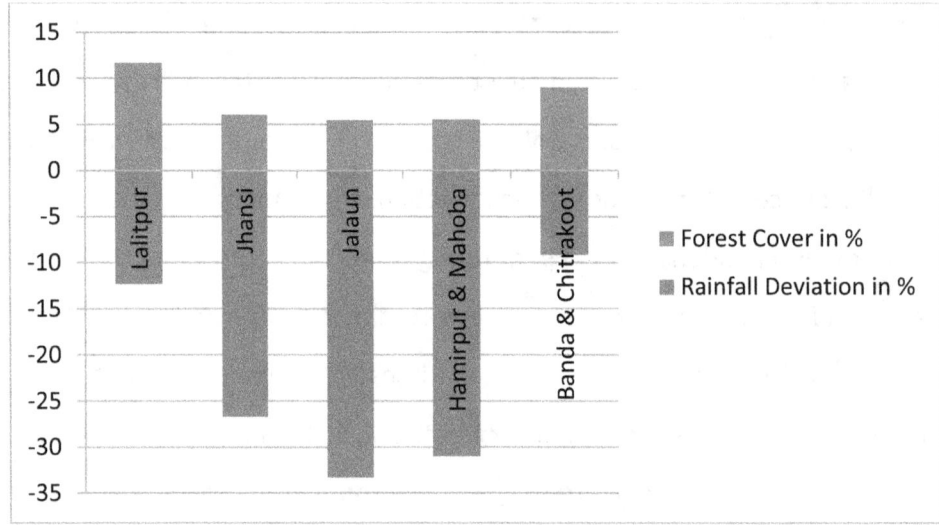

Fig. 4 Forest cover and deviation in rainfall

9.3 Imbalance in Development of Groundwater

Groundwater, which fetches the primary needs of different sectors in the state, faces multiple problems across the state, as depicted in Table 9.5[47] and figure 5. Table 9.5 shows that groundwater faces a variety of problems i.e. lowering of groundwater level, overexploitation, groundwater quality, shallow groundwater level, waterlogging, and groundwater scarcity in various districts. In some districts, groundwater problems are not adequately reflected.

Table 9.5 Groundwater problems across the State of UP

S.no.	Prominent groundwater problems	Name of affected districted
1	2	3
1.	Groundwater level lowering, over-exploitation	Saharanpur, Shamli, Muzaffarnagar, Amroha, Sambhal, Badaun, Meerut, Bagpat, Rampur, Bulandshahar, Aligarh, Hathras, Etah, Mainpuri, Kannauj, Varanasi
2.	Groundwater level lowering, over-exploitation, groundwater quality	Bijnor, Muradabad, Sambhal, Bareilly, Gaziabad, G.B. Nagar, Hapur, Mathura, Agra, Firozabad, Kanpur Nagar, Fatehpur, Pratapgarh, Jaunpur, S.R.Nagar,

		Mirzapur, Lucknow, Faizabad (Ayodhya), Ambedkarnagar, Gazipur
3.	Groundwater level lowering, groundwater quality, shallow groundwater, waterlogging	Raibareli, Sultanpur
4.	Groundwater level lowering, over-exploitation, Scarcity	Jhansi, Mahoba, Banda, Chitrakoot, Sonbhadra
5.	Groundwater level lowering, over-exploitation, groundwater quality, Scarcity	Allahabad (Prayagraj)
6.	Groundwater quality, shallow groundwater, waterlogging	Lakhimpur Khiri, Sitapur, Bahraich, Barabanki, Unnao, Gonda, Balamarpur, Sidharthnagar, Basti, Gorakhpur
7.	Groundwater quality	Pilibhit, Sant kabir Nagar, Balia, Chandauli
8.	Shallow groundwater, waterlogging	Sravasti, Maharajganj, Kushinagar
9.	Groundwater scarcity	Lalitpur, Hamirpur
10.	Problem not adequately reflected	Jalaun, Auraiya, Etawah, Kanpur Dehat, Farrukhabad, Hardoi

Fig. 5 Groundwater problems across the state

9.3.1 Declining Groundwater Level

As discussed earlier, groundwater resource is heavily under pressure in the state. There is overdependence on groundwater to meet the ends with regards to irrigation, industry, domestic and drinking water, etc. About 80 % or more needs of these sectors are fulfilled through groundwater resources, causing over-abstraction and abuse of groundwater. This is ultimately resulting in a heavy toll on groundwater, causing the widespread decline of groundwater levels in rural as well as urban areas.

The yearly decline of 20 cm per year or more is considered critical. Blockwise analysis of pre-monsoon groundwater levels from the year 2009 to the year 2018 shows that 287 blocks are affected with a significant decline of more than 20 cm per year. Out of these, 100 blocks are showing a critical yearly decline of 50 cm or more. Overall, 572 blocks are showing a yearly decline from 01 cm to more than 60 cm as shown in Table 9.6[48].

However, in 248 blocks, either a stable or rising trend in groundwater level has been observed. These blocks need to be evaluated for finding the problem of subsurface water-logging. If these blocks are affected by very shallow water levels or are waterlogged, then other management interventions would be required.

Table 9.6 Blocks Affected with yearly groundwater level decline (Based on pre-monsoon groundwater level data from 2009 to 2018)

Yearly decline range (cm.)	Blocks affected(nos.)
1 - 10	147
>10 - 20	138
>20 - 30	83
>30 - 40	59
>40 - 50	45
>50 - 60	23
> 60	77
Sub Total	572
No Decline/rising trend	248
Total	820

Almost all the prominent urban centers like Lucknow, Kanpur, Meerut, Ghaziabad, Agra, Noida, Varanasi, etc., are severely affected by groundwater depletion. In these areas, groundwater is likely to become a critically scarce resource, as the mining of static groundwater reserves has already started, which is a serious issue and needs urgent attention. Groundwater levels are being monitored in urban areas of the state since 2006-07 through the network of hydrograph stations installed in urban areas. Based on the last 10-12 years of groundwater level data of different hydrograph stations in urban areas, an alarming condition of rapidly declining water levels has been observed. In 22 prominent cities, the groundwater situation has become extremely critical. The water level is declining at a rate of 0.5 m per year to more than 01m every year in cities like Lucknow, Kanpur, Agra, Meerut, Noida, and Ghaziabad. Based on the past trend of groundwater level decline in some cities, the average yearly decline for prominent cities is shown in Table 9.7[5] and figure 6.

Table 9.7.Groundwater level decline in some of the prominent cities of UP

S.no.	Name of the city	The yearly average decline in groundwater level (cm/year)
1	2	3
1.	Meerut	91
2.	Gaziabad	79
3.	G.B.Nagar (Noida)	76
4.	Lucknow	70
5.	Varanasi	68
6.	Kanpur	65
7.	Allahabad	62
8.	Muzaffarnagar	49
9.	Agra	45
10.	Jaunpur	37

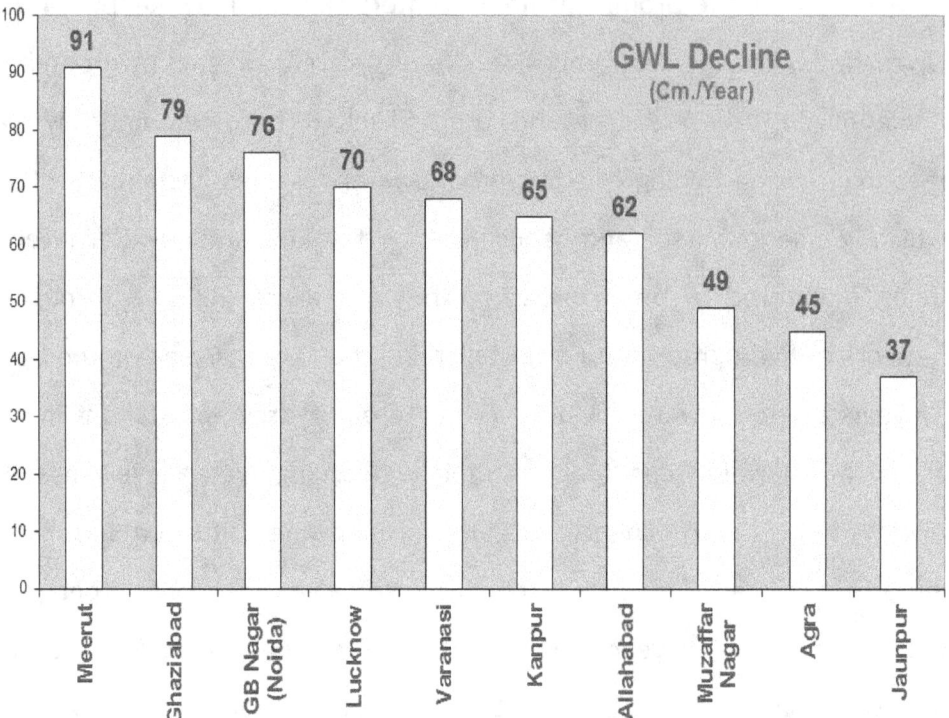

Fig.6 Groundwater level decline in some prominent cities

With the rapid decline of groundwater levels and excessive depletion of aquifers, it would be challenging to ensure future drinking water supplies in most of the urban areas, because it is expected that urban annual water supply would reach a level of about 2400 billion liters or more by the year 2025, while demands of other sectors will also be rising at a faster pace.

Overdependence on groundwater is resulting in the abstraction of fossil or non-renewable groundwater, which is not an active part of the hydrological cycle. This will have grave environmental implications in the future as if this continues; groundwater level may go beyond critical limits of 45 m vanishing natural vegetation which may create severe environmental degradation. Thus, overdependence on groundwater poses a significant challenge to water management in the state.

9.3.2 Shallow Groundwater Level/Water Logging

In canal command areas, crop productivity is being adversely affected due to subsurface waterlogged conditions, especially, in areas where groundwater levels are very shallow (0 m to 3 m). As already reported this area is about 5.689 million hectares. Excess groundwater in these areas can be used to fulfill the needs of scarce areas.

9.3.3 Pollution Hazards

The problem of groundwater pollution is reported from various districts of the state. Potable drinking water and safe irrigation water supplies in many urban and rural areas are severely affected due to chemical, metallic, and bacteriological pollution in groundwater sources. The quality scenario of groundwater resources in the state of UP has significantly changed over the last 2–3 decades. The findings of various quality studies reveal that critical situations related to groundwater quality & pollution hazards are emerging very fast in different parts of the state, which were once known as safe groundwater areas.

Millions of hand pumps & tube wells have been installed all over the state to provide safe & potable water supplies for various uses. The quality reports have shown that the groundwater supplies are now at risk of contamination. The reports have indicated that the groundwater quality in several areas has deteriorated to alarming levels. As a result, human health is now exposed to critical diseases like fluorosis, arsenicosis, heavy metal-related toxicity, and gastro-intentional problem. Various studies on groundwater quality have been taken-up in different parts of the state by different agencies and institutions. These studies indicate that the dynamic shallow groundwater in different areas is variously affected by different quality hazards however, the magnitude of groundwater pollution varies from area to area.

The sources of groundwater contamination are numerous and diverse that includes point and non-point (dispersed) sources. Contamination from point sources is extremely variable with distance and depth and is difficult to characterize from a regional perspective. Likewise, contamination from non-point sources is relatively local and varies with depth. While natural chemicals like arsenic, fluoride, iron are associated with geological formations, the studies have pointed out that the chemical contaminants affecting the quality of groundwater may be of anthropogenic or geo-genic origin; however, the scientific genesis of various contaminants is not yet established [49]. An overview of the districts affected by different groundwater contamination/pollution is summarized in Table 9.8[50].

Table 9.8 Overview of quality affected districts in UP

Contaminants	Districts affected (in part)	Number of districts
Salinity (EC>3000μS/cm at 25°C)	Agra, Aligarh, Firozabad, Ghaziabad, G.B. Nagar, Hamirpur, Kasganj, Kanpur Nagar, Mathura, Raebareli, Unnao, Lucknow, Pratapgarh, Jaunpur	14
Fluoride (>1.5 mg/l)	Agra, Aligarh, Etawah, Firozabad, Hamirpur, Jaunpur, Kasganj, Lucknow, Mathura, Mainpuri, Mau, Raibareli, Unnao, Mirzapur, Sonbhadra, Varanasi	16
Iron (>1.0 mg/l)	Agra, Azamgarh, Ballia, Balrampur, Budaun, Bulandshahar, Chandauli, Etawah, Fatehpur, Firozabad, Ghazipur, Gonda, G.B. Nagar, Ghaziabad, Jaunpur, Kanpur Nagar, Kanpur Dehat, Kannauj, Lalitpur, Lakhimpur Kheri, Hardoi, Mainpuri, Mathura, Mau, Mirzapur, Sant Ravidas Nagar, Siddarthnagar, Varanasi, Unnao	29
Lead (above 0.01 mg/l)	Muzzafarnagar, Mathura, Moradabad, Prayagraj, Bhadohi, Ghaziabad, Jaunpur, Kanpur Nagar, Raibareli, Sonbhadra	10
Cadmium (above .003 mg/l)	Varanasi City, Unnao	2

Nitrate (>45 mg/l)	Agra, Aligarh, Prayagraj, Ambedkar Nagar, Auraiya, Azamgarh, Budaun, Bagpat, Balrampur, Banda, Barabanki, Bareilly, Basti, Bijnor, Districts affected (in part) Bulandshahar, Chitrakoot, Etawah, Etawah, Fatehpur, Firozabad, G.B. Nagar, Ghaziabad, Ghazipur, Hamirpur, Hardoi, Hathras, Jaunpur, Jhansi, Kannauj, Kanpur Dehat, Lakhimpur Kheri, Lucknow, Mahoba, Mathura, Meerut, Mau, Moradabad, Muzaffarnagar, Mirzapur, Raebareli, Rampur, Sant Ravidas Nagar, Shahjahanpur, Sitapur, Sonbhadra, Sultanpur, Shravasti, Sidharthnagar, Unnao.	49
Arsenic (above 0.01 mg/l)	Ambedkar Nagar, Ayodhya, Bahraich, Ballia, Balrampur, Barabanki, Bareilly, Basti, Baghpat, Bijnor, Budaun, Chandauli, Ghazipur, Gonda, Gorakhpur, Kanpur Nagar, Lakhimpur Kheri, Lucknow City, Meerut, Mirzapur Moradabad, Pilibhit, Raebareli, Sant Kabir Nagar, Sahjahanpur, Siddarthnagar, Sitapur, Sant Ravidas Nagar, Unnao	29
Chromium (Above 0.05 mg/l)	Kanpur Nagar, Unnao, Ghaziabad, Varanasi	4
Manganese, Zinc, Nickel, Copper	High levels of these toxic metals are reported from various industrial and urban areas.	

Since groundwater quality analysis is done in isolation, an overall picture of groundwater pollution in the state is not known. Overall groundwater quality zones in the state are not yet mapped and delineated.

Surface water sources also face serious pollution hazards and 70 % of surface water sources are contaminated [53]. All the major rivers are highly polluted as they have become dumping grounds with substantial quantities of contaminants and disposal of untreated effluents from numerous sources along their course. The major factors are industrial untreated effluents, domestic sewage, agricultural runoff, indiscriminate use of polythene, etc. They also face the problem of environmental flow or minimum flow in the lean season due to reduced base flow on account of declining groundwater levels.

9.4 Dealing Groundwater and Surface Water Separately

Groundwater and surface water are part of the same hydrological cycle and cannot be separated. These are complementary to each other. Unfortunately, in the state as well as in the country, they have been operated separately. Conjunctive use of surface water and groundwater has been a popular term for many decades but not adopted at the planning and implementation stage. Nor there is any law to enforce conjunctive use. How to bring changes in the process of planning and implementation to facilitate conjunctive use is a serious challenge.

9.5 Over Dependence on Diesel Energy

As discussed earlier the majority of irrigation water supplies in the country are dependent on electric energy while in the state of UP, they are dependent on diesel energy. Diesel energy, apart from having environmental complications is costly, making agriculture uneconomical for small and marginal farmers.

9.6 Underutilization of Created Irrigation Potential

As stated earlier, about 26 % to 32 % of created irrigation potential is not utilized. This is a considerable percentage of underutilization, and every effort should be made to reduce this. This is a real challenge the state is facing for many years and needs serious efforts.

9.7 Lack of Integrated Planning

NWP, as well as SWP, talk about basin-level planning to develop and manage water resources in an integrated manner, as surface water and groundwater belong to the same resource pool. Revised National Water Policy (2007) recommends integrated and coordinated development of groundwater and surface water resources and their conjunctive use right from the project planning stage and should be an integrated part of project implementation. It also calls for the creation of Basin Authorities.

But, despite policy commitments, there is no organization yet in the state for integrated planning and management of water resources. Growing scarcity and competition for water and overexploitation of water are posing new challenges such as: how to allocate water among competing users, how to improve the productivity of water at farm, system, and basin level, how to enforce conjunctive use, how to control overexploitation of water, etc. To answer these challenges Integrated Water Resource Management (IWRM), which implies that water is treated as a resource and economic good with a wide range of usage (of which irrigation is only one) is being advocated. IWRM can benefit various members and sectors of society. The institutions needed to implement IWRM do not yet exist in the state. Present institutions/departments were created in an era when water was considered plentiful. They deal with water resources in a fragmented manner. The allocation of water among sectors is somewhat arbitrary. The irrigation department is not well informed on groundwater usages within their command areas. Department does not coordinate its activities with other agencies to manage the side effects of irrigation development, including damage to the environment and threats to human health. The same is the case with urban and industrial development authorities/agencies. They generally do not share their future development plans with water resource departments. Similarly, the coordination between the Agriculture Department and Irrigation Department lacks water management issues. Implementation of IWRM requires new management skills, the reform of old institutions, and, most certainly, the creation of new institutions with different cultures and commitments [51].

Though recently state government has taken a positive decision by bringing all water-related departments under the newly created ministry "Jalshakti", at a lower level nothing has changed. Seeing the seriousness of the issue this is a measure too late and too little.

9.8 Low Water Efficiencies

The majority of farmers in the state use open channels and flood irrigation methods to irrigate their fields, resulting in very low water use efficiencies; around 40 % for surface water, and 70 % for groundwater in the state. As discussed earlier, only 0.25 % of minor irrigation structures have micro irrigation systems for water distribution. Agriculture is the primary user of water in the state, and about 85 % of water is consumed to irrigate crops. So, even a small improvement in water use efficiency in the agriculture sector may save a lot of water.

Thus, the state faces a major challenge to increase the area under micro-irrigation. Though of late, farmers of the state have started to shift from open channel to pipe irrigation, adoption of micro-irrigation is not showing the desired results despite heavy subsidy (up to 90 %) for installation of these systems. Some of the reasons may be the following:-

1)   Fragmentation of land holdings

With the ever-increasing population, land holdings in the state are also fragmenting, causing a size reduction. In the year 1995-96, total land holdings in the state excluding the districts of hilly regions or present Uttarakhand State were 18.329 million of which 15.642 million were up to 01 hectares (marginal framers) and 3.004 million were between 01 to 02 hectares (small farmers) [23]. In the year 2010-11, the number of landholdings increased to

23.325 million of which 18.532 million were up to 01 hectares and 3.035 million were between 01 to 02 hectares [16]. Thus significant increase was observed in the number of marginal farmers having a land size below 01 hectares.

The number of operational land holdings by size in UP and some states where adoption of micro-irrigation is comparatively better and in the country are shown in table 9.9[16].

Table 9.9.Number of operational land holdings by size and its percentage   distribution (2010-11)

Size group	Operational land holdings in million and percentage distribution			
	Uttar Pradesh	Maharashtra	Karnataka	India
1	2	3	4	5
Less than 01 hectare	18.532 (79.45)	6.709 (48.97)	3.848 (49.13)	92.826 (67.10)
01-02 hectares	3.035 (13.01)	4.052 (25.58)	2.138 (27.30)	24.779 (17.91)
02 to 04 hectares	1.334 (5.72)	2.159 (15.76)	1.276 (16.18)	13.896 (10.04)
04 to 10 hectares	0.399 (1.71)	0.710 (5.187)	0.511 (6.52)	5.875 (4.25)
10 hectares and above	0.025 (0.11)	0.068 (0.50)	0.067 (0.86)	0.973 (0.70)

Note: Figures in the brackets are percentages

Table 9.9 suggests that the percentage of operational land holdings below 01 hectares in the state is quite high as compared to the national average and figures of the States of Maharashtra and Karnataka. About 80 % of farmers own less than 01 hectare, which is not an economic farm size for capital-intensive agriculture. Drip and sprinkler and other modern technologies remain out of reach of these farmers as these are capital intensive. In these circumstances, micro-irrigation schemes in the present form are not suitable for the state.

2)   **Agricultural practices and crop rotation**

Another reason may be crop rotation and agricultural practices. Major crops in the state are rice, wheat, potato, and sugarcane. With present crop rotation and agricultural practices, micro-irrigation is not economically viable for these crops. These systems can only be feasible in these crops if current agricultural practices of growing these crops are changed altogether. Otherwise, the area under micro-irrigation can be increased by adopting crop rotations suitable for these systems.

3)   **Lack of capacity building**

In the present scheme of sprinkler/drip irrigation, the responsibility of imparting training to the farmers to use these systems properly is on the shoulders of manufacturers who have installed the system. Generally, these manufacturers do not have sufficient manpower to train the farmers. This is an unmonitored area, and nobody is serious about it. The result is that the farmers are not provided technical information such as discharge and pressure ratings of sprinklers/drippers, operational hours required for different crops, layout, and shifting of lateral lines in the sprinkler system, etc. Due to lack of knowledge, farmers are not able to use these systems properly, creating the perception that these systems are not useful or require more diesel or do not provide required water, etc.

9.9 Lack of Effective Unitary Law and Poor Implementation of Existing Laws

Many problems associated with water development and management are not technical but managerial and administrative ones including economic and legal ones and problems related to knowledge, ownership, and regulation. There is a battery of Central and State laws. Some of them have become obsolete in the present context, and some of them are poorly or improperly implemented, virtually making many of them less effective. A brief description of some important acts is the following:-

The Water (Prevention and Control of Pollution) Act, 1994

This act provides for the prevention and control of water pollution and the maintenance or restoration of the wholesomeness of water. As such all human activities having a bearing on water quality are covered by this act. But for practical reasons prioritization of polluting activities is done and industrial pollution is assigned the highest priority.

Water Cess ACT, 1977

The main purpose of this act is to levy and collect cess on water consumed by a certain category of industry.

The Environmental (Protection) Act, 1986

This is an umbrella act for the protection and improvement of the environment and matters connected with it. The government of India has the powers to take all measures as it is deemed necessary or expedient for protecting and improving the environment and preventing, controlling, abating environmental pollution.

Central Ground Water Authority to regulate indiscriminate use of groundwater has been constituted under this act.

Indian Easement Act, 1982

This act links groundwater ownership to land ownership and this legal provision has remained intact since then. Though in landmark Coco-Cola case on the issue of excessive exploitation of groundwater Hon'ble High Court of Kerla observed that the State is the trustee of all-natural resources which are by nature meant for public use and enjoyment and the State as trustee is under a legal duty to protect the natural resources and these resources meant for the public cannot be converted into private ownership but did not uphold the decision of the Village Authority to ban the extraction of groundwater, instead allowed the extraction in the proportion of land owned by the Coco Cola company.

Northern India Land and Drainage Act, 1873

U P Minor Irrigation Act, 1924

Drainage Act, 1880

Irrigation Act, 1876

Constitution 73rd Amendment, Act, 1992

Among other responsibilities, minor irrigation, water management, and watershed development were also included in the responsibilities of Panchayats.

U P Soil and Water Conservation Act, 1963

The U P Participatory Irrigation Management Act, 2009

This act provides for empowering the water users' association to play their role as effective instruments of participatory irrigation management and for matters connected therewith or incidental thereto. Water users' associations have been given an effective role for equitable distribution of water and its efficient and optimum use, operation and maintenance of irrigation and drainage systems, promotion of conjunctive use of surface water and groundwater, command area development, assessment, and recovery of water charges and protection of

environment and ecology.
The U P Water Management and Regulatory Commission Act, 2014 This act provides for the establishment of Uttar Pradesh Water Management and Regulatory Commission to regulate and recommend the tariff for water used for agriculture, industrial, drinking, power, and other purposes and also for levying cess on land benefited by flood protection and drainage, etc.
The U P Ground Water (Management and Regulation) Act, 2019 This act provides for protecting, conserving, controlling, and regulating groundwater especially in stressed areas, and mainly regulates industrial, commercial, and infrastructural, and bulk users of groundwater only in notified and non-notified areas. Its penal provisions do not apply to domestic and agricultural users of groundwater.

According to the constitution of the country, water is a state subject. NWP provides that it is a National Resource and should be treated accordingly. Easement Act, which is a central act, provides that the landowner is the owner of water beneath his land. Environmental Protection Act, which is also a central act, provides that water is a part of the environment. The honorable Supreme Court of India has given the ruling that under the provisions of this act, rules can be framed to regulate and manage water, and under the direction of the court, Central Ground Water Authority was established having jurisdiction all over the country. Almost every state has its irrigation and drainage acts. Many states, on the advice of the Government of India, passed acts to control and manage groundwater. The state of UP also passed the Ground Water (Management and Regulation) Act, 2019. The state already had Uttar Pradesh Water Management and Regulatory Commission Act, 2014. So there is total confusion about the legal status of water and its ownership. If it is a national resource, it should be governed and regulated by national laws. When it is a National Resource, it becomes a common property resource, how the owner of the land can be the owner of groundwater beneath it?

Moreover, the implementation of existing laws is so weak that these remain only on paper and are not able to make any difference with regards to regulation and management of groundwater. Even the Central Ground Water Authority is unable to make any impact and the honorable National Green

Tribunal has to interfere time and time again. Honorable courts cannot do the functions of the executive.

Generally, these acts do not address large classes of management needs and are primarily designed as a mechanism to impose restrictions. Waterlogging, water quality, water pollution, end-use efficiency, allocation, and environmental considerations also represent equally important management challenges and require attention. Apart from this, most of these acts are implemented by government agencies and are driven by the mentality of the British Raj. Stakeholder/beneficiary/people's participation in the implementation of these acts is negligible making them less effective or acceptable. For enforcement, most of these acts include search and seizure provisions with substantial fines and imprisonment; these regulatory approaches are unlikely to be successful. Such laws are inherently difficult to enforce, seeing the small landholdings and inadequate administrative setup. Besides, the absence of an effective role for the local population in decision-making may intensify opposition, and opportunities will be missed for the development of management approaches reflecting local interest and considerations [51, 52].

9.10 Water Scarcity

The whole country is now classified as water-stressed, and so is the state of U P. Approximately 820 million people of India in twelve river basins across the country have per capita availability of water close to 1000 cubic meters or less which is the official threshold of water scarcity. Two major crops of the country i.e., wheat and rice are being affected by water-related issues. About 74 % of the area under wheat and 65 % of the area under rice faces significant levels of water scarcity. India is also losing large quantities of water through water loss on account of the export of water-intensive crops. Overall, the water demand of the country is expected to exceed the water supply two-fold by 2050. Annual per capita water availability in India is expected to reduce to 1140 cubic meters by 2050, close to water scarcity thresh hold of 1000 cubic meters. If the present scenario continues, 6 % of India's GDP will be lost by 2050 [53].

In the state of UP also, wheat and rice are the major crops. With diminishing rainfall, surface flows are reducing, and a declining trend in groundwater level has also been observed in 572 blocks. Out of 820 assessed blocks, 280 are categorized as overexploited/critical/semi-critical. Its major river basins such as Gomti, Ramganga, etc. are already water short. It is home to more than 200 million people or about 17 % of the population of India. Required food production for its projected population by 2050 will be about 86 to 109 million metric tons, for which about 105 to 143 BCM of water would be required. To achieve this, the state will have to enhance net irrigated area from present 14.33 million hectares to 18.2 to19.2 million hectares and crop intensity to 180 to 200% from present 157.5% [54, 16].

Thus, achieving food security for its rising population will be a daunting task, and water scarcity will make this task tougher. The Composite Water Management Index (CWMI) of the state is only 38.7 % (the reference year 2017-18), and no significant improvement is observed in the previous two years [53]. So the state will be facing significant water scarcity in the future.

9.11 Water Pricing

Water pricing is another ignored aspect of water management in the state. Successive governments opted policy of free water to the farmers as a populist measure, which is costing heavily on water resources. It is widely accepted that the free availability of any commodity amounts to its misuse, wastage, and abuse. Israel, a tiny country having very scarce water resources, acknowledged all over the world for its water management efforts, adopted the philosophy that free water is more costly and priced it appropriately. Proper pricing has the potential to change water use patterns. Israel used this concept effectively by pricing treated water, less as compared to freshwater for irrigation [55]. Pricing water is one of the main functions of Uttar Pradesh

Water Management and Regulatory Commission but it has not been able to do it yet as it lacks infrastructure, and its powers are advisory.

10. FUTURE STRATEGIES/INTERVENTIONS

With diminishing and erratic rainfall, the situation in the state has deteriorated rapidly after the year 2000, and climate change has further aggravated the situation. The state has taken many policy initiatives and started to reform the governance of water, though, at a meager pace. The state has to increase the pace of these initiatives, try innovative and area-specific out of box solutions along with traditional approaches to counter the new challenges the water sector faces today. The situation has now reached a stage where only the traditional approach and thinking will not be of any help. For example, after the year 2000, the state adopted the strategy to construct as many traditional rainwater harvesting structures as possible under different ongoing schemes to manage the situation and increase water availability, but this has not yielded the desired results as nature is changing more rapidly than the efforts put in. Moreover, when rainfall is diminishing, merely supply-side management initiatives are not helpful. In such a situation, more emphasis is required on demand-side initiatives. Similarly, there are many laws for water, but their implementation is poor. The absence of a concrete regulatory regime and robust mechanism of implementation leads to mismanagement of water resources. There are many other examples also. Some of the future strategies/interventions are discussed below.

10.1 Increase Forest Cover

Presently forest cover in the state is stagnating at around 6% of the area, which is far less than the standard one-third or about 33 % of the area. This is not evenly distributed also within the state. With traditional measures and programs of forest management, increasing forest cover up to one-third of the area seems an uphill task. Planting 200 or 220 million saplings every year is not working. Instead, there is a need to develop new forests. It is suggested to develop a forest of 100 hectares or more in each village, as was the traditional practice in the past. The latest technologies to develop forest rapidly as adopted by Japan in small patches may be an innovative idea. This approach will also lead to even distribution of forest cover in the state. Special laws can be promulgated for this, and if the land is not available, it can be acquired. This may be done with the active participation of the local population. To implement this approach, a separate Authority or Special Purpose Vehicle may be constituted.

10.2 Improve Per Unit Productivity of Water

This involves improving the water use efficiency by promoting the use of sprinkler/drip irrigation in place of flood irrigation, which is the norm in the state. Uttar Pradesh is the largest agricultural producer in India but has negligible micro-irrigation coverage, a worrying fact. It is suggested to review the ongoing micro-irrigation schemes and ascertain the reason behind the negligible coverage despite a high level of subsidy. Some of the probable reasons are already discussed in the previous sections. In light of these, it will be appropriate to design group schemes such as horticulture complexes, etc. for sprinkler and drip irrigation in the state capable of changing the existing crop rotation and agricultural practices.

Other water-saving measures, such as the use of water retention polymers, sub-surface irrigation methods, smart water metering in domestic piped water supply should also be tried on priority. The domestic water supply is not metered in the state and surprisingly there is no sincere effort in this direction. This shows that water is not being recognized as an economic good by the system.

10.3 Convert Waste Water into Water

With declining and erratic rainfall, this reliable source of water supply is now no more reliable. With growing demand state has to search for alternative sources of water supply. Wastewater/sewage water may be one such source as it is available round the year in the same quantity. Israel over several decades has already done it by building national wastewater infrastructure to make the use of wastewater, especially for irrigation, and presently uses 85 % of its wastewater [55]. Wastewater generated from large urban areas of the state may be converted into the water to develop a new source of water supply for which infrastructure to collect and filter wastewater, store and transport it to the scarce areas will have to be developed. This requires new thinking, a new vision, a new organization, and new investment. The present system of draining treated sewage water from sewage treatment plants into the river should be relooked as this water after further treatment can be stored and transported for irrigation and other needs.

Accurate and consolidated information regarding generated wastewater is not available for the State of Uttar Pradesh. However, it is also true that it is a large state that has a 45 million urban population approximately and generates large quantities of wastewater daily. The average consumption of freshwater is 135 to 150 liters per capita per day (lpcd). If we take daily per capita freshwater consumption as 135 liters, per capita wastewater generated is 90% or about 121 lpcd. However, it is the most conservative estimate; in actual practice, it will be much more. For a population of 45 million, the amount of generated wastewater comes to be 1987.42 million cubic meters (MCM) per year. It does not include wastewater generated from industries, generally 50 to 60 % of domestic wastewater. Total wastewater generated will be around 3080 MCM, which is a huge quantity, most reliable, available round the year in the same amount. It will increase with an increasing population, which can be used to supplement the irrigation water demand of the parched regions like Bundelkhand. Instead of limiting the strategy to rainwater harvesting, it is now the right time to develop the integrated facilities to retrieve, treat, store and transport the wastewater so that in case of emergencies, it can be used to supplement the demand of parched regions [57].

10.4 Review Legal Status of Water and Present Water Laws and Enact a Strong Unitary Law

(i) Many water laws are having overlapping provisions creating more confusion than facilitating the management of water. For example, the recently promulgated Ground Water (Management and Regulation) Act, 2019, has strict penalties for groundwater pollution. At the same time, groundwater pollution is also covered under the central law "Water (Prevention and Control of Pollution) Act, 1974. Central Ground Water Authority, constituted under Environmental Protection Act, 1986, has powers to notify groundwater stressed areas while State Ground Water Management and Regulatory Authority constituted under The U P Ground Water (Management and Regulation) Act, 2019 also has powers to notify groundwater stressed areas in the state. By nature and according to NWP and SWP, water is a common property resource, but the Easement Act, 1986, gives ownership of groundwater to private persons. The newly promulgated UP Ground Water (Management and Regulation) Act, 2019, provides to charge a fee for groundwater extraction from industries while Water Cess Act, 1977, also has this provision. So there is a need to review all the existing laws and promulgate a unitary law that is strong and effective.

(ii) Existing water laws are needed to give more teeth to manage water resources properly. For example, the provisions of the Uttar Pradesh Water Management and Regulatory Commission Act, 2014, are advisory. Similarly, Uttar Pradesh Ground Water (Management and Regulation) Act, 2019, is only for controlling commercial/bulk groundwater extraction while major misuse is in the agriculture sector which uses 85 % of the water in the state. Though in this act, there are some provisions for agriculture and domestic users of groundwater, penal provisions do not apply to these users. So in case, these users contravene the provisions of this act, there is no tool to enforce them on these users. Virtually about 98 % of the groundwater users are not regulated through this act. Similarly, the constitution of Water Users Associations under

Uttar Pradesh Participatory Irrigation Management Act, 2009, is very slow as provisions of this act are not mandatory. Timely creation of WAUs and regular elections are crucial, without which this act becomes ineffective.

(iii) State also needs to manage its international export of virtual water and availability. To do this, the crop area of the different crops will be required to be regulated for which neither there is any law nor any provision in the existing laws in the state. It is discouraging to see that farmers in the parched region of Bundelkhand producing water-guzzling crops like mentha or sugarcane or paddy or wheat. It is now necessary to enact the law on this important aspect to ensure that crop production patterns are aligned to district-level/block-level water availability.

(iv) The Easement Act, 1882, which is a central act, gives ownership of groundwater beneath a piece of land to the landowner. When NWP and SWP recognize water as a common property resource, how an individual can be the owner of it. The central government must be requested to abolish this provision. National Water Policy recognizes water as a national resource. Then how a national resource can be a state subject? Now the time has come to have a unified central law to manage water resources more effectively. The state may request the Government of India to reconsider the constitutional provisions regarding the legal status of water.

(v) (v) Provisions of the present acts are generally regulatory, having nil involvement of the affected population, and implemented by government-controlled agencies. However, for resource management, there is a need to create basins/groundwater basins or groundwater districts controlled by people's organizations having complete authority on administrative, financial, and technical management. These organizations should be completely independent, having powers to impose restrictions, levy water development charges, float their bond for financial resources, recruit technical persons as required, etc. Similar authorities are also required for river basins.

(vi) Pricing of water is another aspect of the water management matrix which needs serious attention. Though fixing tariffs for different water usage is an essential function of Uttar Pradesh Water Management and Regulatory Commission, no progress is seen in this regard as it lacks infrastructure and powers.

10.5 Simultaneous or Conjunctive Use of Surface and Groundwater

Conjunctive use or simultaneous operation of surface water and groundwater had been the much-talked term in past decades, but it is not an integral part of the planning process yet. Implementation of this concept can check the overexploitation of groundwater, resulting in better management of water resources and fetch the needs of water-scarce areas. There may be numerous extensions of this approach.

Water Users Associations in the United States of America (USA) are transferring large quantities of water from surface reservoirs just before the start of the rainy season into the depleted aquifers through pressure head injection wells recharging the aquifers as well as creating extra space to store more rainwater. This is equivalent to constructing many new dams. Apart from this, the recharged water is transported hundreds of kilometers down the stream through aquifers, where farmers retrieve it for use at the time of need [56]. This also reduces the cost of water conveyance and solves the problem of pollution and evaporation as encountered in open canals. These types of out-of-box solutions are needed in the state too. Of course, they require huge investment and capacity enhancement, but these are the need of the hour. To begin with, some pilot projects can be taken up.

Another extension of this approach may be in waterlogged areas. As stated earlier, the state has a sizable waterlogged area which means there is plenty of unutilized groundwater. Through a combination of surface and subsurface field drains, the water level in these areas can be lowered, and excess water can be collected in large sump wells and pumped to water-scarce areas through piped water carriers. These ideas and technologies are not tried in the state but have the potential to change the water management scenario in the state and deserve to be tried.

It has been observed the world over that conjunctive use is by chance not by planning. So, to realize this approach integrated planning or conjunctive planning and implementation of surface water and groundwater projects is a prerequisite for which creation of state-level planning office and basin level organizations is the first step, but despite policy level commitments, they are nowhere in the picture. So, transforming present water organizations and creating new basin-level organizations should be accorded the highest priority. This is a difficult task but possible with the intervention at the apex level of the government. Making conjunctive use mandatory through law may be another option as SWP also provides for it.

10.6 Reconciliation of the Figures of Created Irrigation Potential and Initiate Measures to Reduce Gap between Potential Created and Utilized

As stated earlier, figures of irrigation potential created are not consistent and much more than the assessed ultimate irrigation potential of the state. The loss of irrigation potential over the years is also not reconciled in these figures. Moreover, the majority of Major and Medium Irrigation Projects are age-old and their irrigation potential was worked out according to the crop water requirements of prevalent crop rotations at that time. Over the years, crop rotations have changed, and now farmers are growing crops having high water requirements. The same is the case with minor irrigation projects. So, it is obvious to reassess the created irrigation potential of each project incorporating loss of irrigation potential. After reconciling the figures of created irrigation potential, the real picture regarding the gap between potential created and potential utilized will emerge. If the gap is significant, the reasons for the gap are to be identified and addressed effectively. Instead of investing in new projects, the topmost priority should be on reducing the gap so that investments already made are properly utilized. It will be appropriate to constitute a working group to address all these issues effectively.

10.7 Create New Organization to Monitor and Control Water Quality

Water quality is another issue that needs immediate attention. About 70 % of surface water is contaminated. The first aquifer in the state is almost contaminated, and with the decline in groundwater level, problems of arsenic and fluoride contamination are also emerging. So there is a necessity to strengthen the water quality monitoring network and check the sources of pollution effectively; otherwise, the sustainable development goal of providing universal access to safe water will be impossible to achieve. It will be appropriate to create a new organization equipped with the required infrastructure and powers. At present different organizations monitor water quality without any coordination and the present infrastructure seems insufficient with regards to the magnitude of the problem.

10.8 Improving Water Supply through Redevelopment of Traditional Water Bodies

As stated earlier, there are plenty of traditional water bodies in the state, most of them in very dilapidated conditions currently. There is a need to map, renovate, protect, and preserve them as water sanctuaries. This will go a long way towards improving water supply, checking falling groundwater levels, and improving base flow into the rivers. Revitalization of the traditional water bodies should also be accompanied by promoting water as the common property resource for the people that need to be protected and enhanced by the community for the benefit of the community. Community-level conservation of water is the key to improve water supply at the local level.

Off late, the state government has started a new scheme for redevelopment and management of ponds (> 01 hectares). This scheme has an innovative feature of "Pani Panchayat" (democratization of water) to ensure public cooperation. The initial response to the scheme was good and there was great public demand, but fund allocation for the scheme is not sufficient. Only INR 480 million are provided per year for the scheme from which about 430 ponds can be redeveloped. Seeing a large number of ponds of one hectare or more, this allocation seems insufficient. It will take many years to redevelop all the ponds. Another issue is about the rights of Pani Panchayat. After initial enthusiasm, members of Pani Panchayat lost interest as they do not have any right over the ponds and their water. These ponds are in the ownership of the revenue department, which gives these ponds on lease for fishing etc. So, to ensure public cooperation, the rights on the pond and its water needs to be handed over to Pani Panchayats.

10.9 Introduce Internet-Based Technological Interventions to Automatize Water Sector

Automatization is an effective way to improve efficiency. Farmers of the state generally irrigate their crops as per traditions. They have little knowledge/information about changing climate and rainfall patterns and what and how to do in changing situations. As the world is trending to modern technologies, it is necessary to trend up the irrigation/agriculture sector also. Unfortunately, the penetration of information technology in this sector is very low. Internet of Things and cloud, in combination with wireless sensor networks, can lead to agriculture modernization and automatize irrigation. Wireless sensor network monitors parameters like temperature, humidity, rainfall, soil moisture, etc., and analyses threshold values to start or stop irrigation or need of fumigation, etc. These modern technologies have great potential to improve efficiency and must be tried at the earliest to prove their efficacy in actual field conditions for their large-scale adoption.

Apart from this, using weather and climate data, specific to the district, suitable models can be developed to forecast future action in advance and communicated to the farmers through a mobile application. In today's rapid technological advancements, no state can afford to be technologically backward.

10.10 Groundwater Recharge/Restoration

Up till now, the state adopted the strategy heavily based on traditional ways of groundwater recharge which are mainly rainwater harvesting and water conservation methods. It has not started any specific recharge project, introducing modern technologies of groundwater recharge such as injection wells, infiltration galleries, radial wells, bore blasting techniques, hydro fracturing, etc. Though the state has 17 % of India's groundwater, it has only mapped 20 % overexploited and critical blocks and constructed no recharge infrastructure [53]. Thus there is a need to give new thinking to groundwater restoration and make an investment in recharge infrastructure. The traditional way of doing things is not going to help in countering the emerging challenges.

Urban areas have different characteristics and need different solutions. Hear, collection areas are rooftops and surface pavements (roads) and open spaces such as parks, etc. While in most of the urban and semi-urban areas, groundwater levels are declining at alarming rates, efforts for restoration of groundwater are limited to mandatory provisions of recharge structure on the plots of 300 square meters or more, multi-storied and government buildings without any robust regulatory mechanism. The result is that most of these structures become defunct after some time for want of cleaning and maintenance.

No serious attempt has been made to introduce the concept of surface pavement rainwater harvesting and recharge or building ponds or reservoirs and divert rainwater into them. Large quantities of rainwater can be collected from surface pavements and diverted to storage structures built either on the surface or underground, and this stored water can be used to supplement the water supply after filtration. Parks can be used to build underground storage structures as new technologies are now available to build such underground structures without disturbing the landscape of the park. These are not difficult tasks provided that organizations and authorities concern stop thinking and planning in traditional ways, rather embrace the change. Further, organizations having the capacity to implement these types of projects are needed.

10.11 Regulation of Groundwater Based Irrigation Schemes

Many districts of Western UP in the state, now, have 90 % or more irrigated areas. Logically, groundwater-based government schemes should have been banned in these districts as now there is no point in giving subsidies to encourage the construction of tube wells or construction of a government tube well in these districts. The state needs to close all government schemes of groundwater development and groundwater extraction activities, whether private or government, except related to drinking water in those districts where the irrigated area is more than 90%. There is no logic to give subsidies or spend government funds for groundwater extraction in these areas, rather society should be encouraged to share the already developed facilities whether government or private and improve water productivity. Financial resources saved from the closure of such schemes can be utilized for demand-side initiatives.

CONCLUSIONS

In its journey of water development, the achievements of the state are remarkable. After the beginning of the plan period, the state could increase the irrigated area to the tune of 450 %, which is no doubt a great achievement. The number of private tube wells, after the advent of irrigation pump sets in the eighties, swelled in the state from 1.8 million to about 3.8 million or an increase of about 220 %. But in its endeavor to develop irrigation facilities at a fast pace, much emphasis was given to groundwater-based irrigation infrastructure, resulting in an imbalance between the development of groundwater resource and surface water resource. More than 80 % of the net irrigated area is attributed to groundwater and 19 % to surface water, while the total dynamic groundwater resource or replenish-able groundwater resource is less than the surface water resource. The same is the case with the domestic and industrial sectors, where dependence on groundwater is more than 80 %. The state needs to make all-out efforts to reduce dependence on groundwater in the coming decades; otherwise declining groundwater levels may lead to serious environmental degradation.

Management of any resource requires reliable, accurate, and timely information. In the absence of these, wrong decisions can be taken or wrong projects can be started. Many discrepancies are observed in the figures of created irrigation potential, waterlogged areas, dynamic groundwater resource assessment, ultimate irrigation potential, etc. For better water management of water resources state needs to reconcile the figures of created irrigation potential, review groundwater assessment methodology, reassess ultimate irrigation potential through surface water and groundwater resources. Further, there is a lag of three to five years in publishing the information/data. This gap needs to be reduced as the entire planning and execution are based on three to five years old information/data.

The state's achievement in increasing food production is also laudable. Since the mid-eighties, the state achieved a growth of 186 % in total food production. Production of major crops like rice, wheat, sugarcane, and potato increased manifold, and the state is now one of the major contributors to the food basket of the country. However, the average yield of two main crops i.e., rice and wheat, are still far below the leading states like Punjab and Haryana.

The state took many policy decisions starting with State Water Policy in the year 1999, followed by Word Bank assisted project to introduce water sector reforms, observance of Ground Water Week, the inclusion of rainwater harvesting in the school curriculum, promulgation of Uttar Pradesh Participatory Irrigation Management Act, 2009, Policy for Ground Water Management, Rain Water Harvesting, and Groundwater Recharge, 2013, Uttar Pradesh Water Management and Regulatory Commission Act, 2014, Uttar Pradesh Ground Water (Management and Regulation) Act, 2019, the establishment of State Water Resource Agency, State Water Resource Data Analyses Centre, Water Management and Regulatory Commission, launching of schemes like Redevelopment and Management of Ponds, National Hydrology Project, Atal Bhujal Yojana, creation of unified Water Ministry, etc., the pace of these initiatives/reforms/schemes is sluggish. Implementation infrastructure for new acts is weak without any role of the affected population, or their provisions are not mandatory, making them practically ineffective. Most of the new schemes also lack human and financial resources.

Climate change impact is surfacing in the form of diminishing and erratic rainfall with increasing frequency of drought. The state is not able to enhance forest cover substantially, which is stagnating around 6 % since the year 2000, far below the standard one-third, i.e. 33 % of the area. The state also faces the problem of declining groundwater levels in a sizeable area, waterlogging, underutilization of created irrigation potential, low water efficiency, water scarcity, integrated planning and implementation, simultaneous operation of groundwater and surface water, poor implementation of existing policies and relevant laws, water quality and lack of initiative about groundwater recharge/ restoration projects and adoption of modern technologies.

However, in its preoccupation with developing water resources, management aspects of water resources were mainly limited to supply-side management. After the year 2000, climate change started showing an adverse impact on water resources, and within two decades, the situation changed dramatically, bringing the state into the water-stressed category. In coming decades, to fetch future water demand of its growing population, the state needs to invest heavily in terms of technology, human resource, organizational change, capacity enhancement etc. to increase micro irrigation coverage, automatize irrigation using internet of things, cloud and sensors, start specific groundwater recharge/restoration projects using state of the art technologies, create infrastructure to implement conjunctive use, convert waste water in to water and use it, reclaim access water from water logged areas to fetch the needs of water scarce areas, redevelop and manage large water bodies, embark upon innovative and new strategies/initiatives to increase forest cover, improve water productivity, promulgate unitary strong water law with proper implementing infrastructure, create reliable, accurate and up to date data base, create effective water quality monitoring and control infrastructure, make system through legislative measures to ensure crop area in line with the availability of water at district/developmental block level, enforce water pricing, enhance allocation for redevelopment and management of large water bodies, promote concept of "Pani Panchayat" by giving them rights on water bodies and its water, make policies for involvement of society in water governance and implementation and developing social leadership at local level.

With the present climate crisis, the state is entering a whole new era. To cope with it, a new mind-set, new ways, new ideas, new means, and a new culture will be required. Whether the future generations will get sufficient and safe water will depend on the speed with which the state takes measures to cope with water scarcity triggered by climate change, otherwise, history may repeat itself, and the present civilization of the state developed along the great Ganga and Yamuna Rivers and their tributaries may vanish like the Indus Valley Civilization. Some of the new ideas are narrated under the heading Future Strategies/ Interventions, but these are not exhaustive and there may be many more innovative ideas.

REFERENCES

1. Government of Uttar Pradesh, State Planning Commission. 2009. *"Draft Annual Plan", 2009-10*, Vol.-I (Part II), 86-87.

2. Government of Uttar Pradesh. 1999. *"State Water Policy", 1999*, 29, 31-33, 39-40, 48-49.

3. Government of India, Central Ground Water Board (NR), Lucknow, and Ground Water Department UP. 2019. *"Dynamic Groundwater Resources of Uttar Pradesh (Reference Year 2017)"*, November, 13, 276, 324.

4. Sinha R. S. and Chaurasia, P. R. 2019. *"Critical Overview of Ground Water Management Policies and Practices: Need to have a Robust Management Mechanism."* Proceedings, All India Seminar on Sustainable Water Management & Conservation, the Institution of Engineers (India), UP State Centre, Lucknow, UP, November 2-3.

5. Government of UP, Ground Water Department. 2017. *"Atlas of Ground Water Resource Maps"*.

6. Government of India, Central Ground Water Board, Ministry of Water Resources, River Development and Ganga Rejuvenation. 2017. *"Report of the Groundwater Resources Estimation Committee (GEC-2015) Methodology"*, October 2017, New Delhi.

7. Irrigation and Water Resource Department UP. *"World Bank Schemes/National Hydrology Project."* Accessed March 2020. http://idup.gov.in/pages/en/topmenu/w.b.-aided-projects/en-national-hydrology-project-nhp.

8 Khan Seraj. 2017. *"Hydrology of Uttar Pradesh",* Central Ground Water Board, Norther Region, Lucknow, Ministry of Water Resources, River Development & Ganga Rejuvenation, Government of India, 48-50, 155-158.

9. Sharma, S.K. 2010. *"Strategies for Ground Water Management in State of Uttar Pradesh",* Book on Sustainable Water Management, First Edition, Lucknow, UP, Connoisseur Publishers, 36.

10. Government of India, Central Ground Water Board. 2019. *"National compilation on Dynamic Groundwater Resources of India, 2017"*, July 2019.

11. Government of India, Ministry of Water Resources. *"Annual Report, 1997-98"*, 18.

12. Government of India, Ministry of Water Resources. *"Annual Report, 2013-14"*, 23-24.

13. Sinha, R. S., Anupam, Kumar A. and Upadhayay, V.K., 2019, *"Automated Ground Water Level Monitoring in Uttar Pradesh-A Way forward under World Bank Project",* proceedings, All India Seminar on Sustainable Water

Management & Conservation, The Institution of Engineers (India), UP State Centre, Lucknow, UP, Nov 2-3.

14. Government of Uttar Pradesh, Ground Water Department. 2019. *"Progress Report (Groundwater Component), UP Water Sector Restructuring Project, Phase-II"*, September.

15. Kapoor S. K. and Khan A. A. 2006. *"Impact of Minor Irrigation Schemes on Agricultural Productivity"*, Souvenir, All India Seminar of Minor Irrigation Works and its Impact on Agriculture Production", the Institution of Engineers (India), UP State Centre and Minor Irrigation Department, UP, Lucknow, UP, January 16-17.

16. Government of Uttar Pradesh, Economics and Statistics Division, State Planning Institute Uttar Pradesh. 2018. *"Statistical Diary, Uttar Pradesh, 2018"*, 4, 129-195. Accessed January 2020. http://updes.up.nic.in

17. Government of UP, Minor Irrigation Department. 2014. "Report of *Fourth Census of Minor Irrigation Schemes (Reference Year 2006-07) in the State of UP"*.

18. Ministry of Water Resources, River   Development and Ganga Rejuvenation, Government of India. 20117. *"Report of Fifth Census of Minor Irrigation Schemes (Reference Year 1013-14)"*. Accessed January 2020. Htttp://www.mowr.gov.in.

19. Chaurasia P. R. and Subhash. 2018. *"A Low-cost Indigenous Intervention which has Revolutionized the Drilling Technology and Changed the Life of Millions of Farmers in the State of Uttar Pradesh, India"*, Journal of Institution of Engineers (India), Series A, Volume 99, Issue 2, 2018. http://doi.org/10.1007/s40030-018-0278-7

20. Michael A.M. 1978. *"Irrigation, Theory, and Practice"*, New Delhi, Vikas Publishing House, 36.

21. Government of India, Ministry of Water Resources, Minor Irrigation Division. 2005. *"Report on 3rd Census of Minor Irrigation Schemes (Reference Year 2000-2001)"*.

22. Indian Water Resource Management Society. 1998. *"Main Document on Five Decades Development of Water Resources in India"*.

23. Government of Uttar Pradesh, Economics and Statistics Division, State Planning Institute Uttar Pradesh. 2000. *"Statistical Diary, Uttar Pradesh, 2000"*, 91-144.

24. Chauhan H. S. and Rai Dhaneshwar. 2006. *"Prevention and Sustainable Reclamation of Water Logged and Degraded Lands in UP"*, Souvenir, All India Seminar on Minor Irrigation Works and its Impact on Agricultural Productivity, The Institution of Engineers, UP State Centre and Minor Irrigation Department UP, Lucknow, India, January 16-17.

25. Government of India, Ministry of Water Resources. *"Detailed Note on Reclamation Measures in Water Logged Areas of Irrigation Command"*. Accessed February 2020. Http://mowr.gov.in.

26. Government of India, Regional Remote Sensing Service Centre, Jabalpur, Indian Space Research Organization and Central Water Commission, New Delhi. 2009. *"Assessment of Waterlogging and Salt and/or Alkaline Affected Soils in the Commands of All Major and Medium Irrigation Projects in the Country Using Satellite Remote Sensing, Country Report"*. Accessed February 2020.   http://www.indiawaterportal.org.

27. Uttar Pradesh Jal Nigam. 2018. *"Water Supply Status Report"*, September".

28. Government of UP, Ground Water Department. 2014. *"Compilation of Policy Decisions and Government Orders on Rain Water Harvesting and Groundwater Recharge in Uttar Pradesh"*, July.

29. Government of UP. 2009. *"The Uttar Pradesh Participatory Irrigation Management Act, 2009"*.

30. Government of UP. 2013. *"Policy for Ground Water Management, Rainwater Harvesting and Ground Water Recharge"*.

31. Government of UP. 2014. *"The Uttar Pradesh Water Management and Regulatory Commission Act, 2014"*.

32. Government of UP. 2019. *"The Uttar Pradesh Ground Water (Management and Regulation) Act, 2019"*.

33. Government of Uttar Pradesh, Ground Water Department UP. 2017. *"Order Regarding Bhujal Sena"*.

34. Government of UP, Minor Irrigation Department. 2017. *"Order Regarding Pani Panchyat"*.

35. Government of UP, Ground Water Department UP, *"National Hydrology Project"*. Accessed March 2020.   http://upgwd.gov.in/StaticPages/World Bank Scheme-hi.aspx

36. Government of UP, Ground Water Department. 2019. *"Guide Lines of Atal Bhujal Yojana"*.

37. Government of UP. *"State Water Resource Agency, UP and State Water Data Analysis Centre"*. Accessed March 2020. www.swaraup.gov.in

38. Government of UP, Irrigation and Water Resources Department UP. 2014. *"Order Regarding Constitution of Water Management and Regulatory Commission"*.

39. Government of UP, Irrigation and Water Resource Department. 2018. *"Order Regarding Appointment of Chairman, Water management and Regulatory Commission"*.

40. Government of UP, Irrigation and Water Resources Department UP. 2018. *"Order Regarding Creation of Namami Gange and Rural Water Supply Department"*.

41. Prakash Bhartendu, Ghosh Shailendra M., Satya, Santosh & Chaurasiya, L.P. 1998. *"Problems and Potential of Bundelkhand with special reference to Water Resource Base"*, Center for Rural Development and Technology, India, Institute of Technology, Delhi and Vigyan Siksha Kendra, Attara (Banda), UP, 2-8.

42. Gupta A.K., Nayar, Sreeja S., Ghosh, Oishance, Singh, Anjali & Dev Sunand. 2014. *"Bundelkhand Draught - A Perspective Analysis and Way Ahead"*, National Institute of Disaster Management, New Delhi 2014, 2-8.

43. Ministry of Earth Sciences, New Delhi, India, Hydro met Division, India Metrological Departments. *"Customized Rainfall Information System"*. Accessed April 2018.   http://hydro.imd.gov.in/hyrometweb/District rainfall

44. Government of India, Forest Survey of India, Ministry of Environment and Forest, *"State of the Forest Report, Uttar Pradesh, 2001 to 2017"*. Accessed April 2018.   www.fsi.nic.in.

45. Global Land Scape Forum Bonn. 2017. *"Precipitation and its Relation to Vegetation"*. December 19-20.

46. Sheir Douglas and Murdiyrse Doniel. 2009. *"How Forest Attract Rain: An Examination of New Hypothesis"*, BioScience, Vol. 59, Issue 4, April 2009, 341-347.

47. Government of U.P., Ground Water Department, *"Groundwater Resources of Uttar Pradesh- An Overview"*, Lucknow, UP, 12.

48. Government of U.P., Ground Water Department, *"Groundwater Data"*. Accessed March 2020.
upwd.gov.in

49. Sinha R.S. 2011. *"Arsenic Toxicity in Ground Water of Uttar Pradesh"*, Technical Report, GWE/1/2011, State Water Resources Data & Analysis Centre and State Water Resources Agency, Lucknow U P, February.

50. Government of India, Central Ground Water Board (NR), Lucknow, *"State Profile, Groundwater Scenario of Uttar Pradesh"*. Accessed March 2020.
cgwb.gov.in

51. Chaurasia, P.R. 2009. *"Legal and Institutional Framework for Groundwater Management – Status and issues"*, Proceedings, State Level Workshop on Groundwater Management in Uttar Pradesh – Challenges, Priorities and

Strategies, State Water Resource Agency, U.P. and State Water Resources Data Analysis Center U.P., Lucknow, India, August, 20-21.

52. Moench, Marcus. 2016. "Drawing Down the Buffer: Science and Politics of Groundwater Management in India." In *Water: Growing Understanding, Emerging Perspective,* edited by Mihir Shah and P.S.Vijayshankar, 112-116, New Delhi: Orient BlackSwan Private Limited.

53. Government of India, NITI Ayog. 2019. *"Composite Water Management Index Report"*. Accessed February 2020. pp 1-31, 208. http://social.niti.gov.in/uploads/sample/water_index_report2.pdf

54. Kumar Ravindra. 2009. *"UP Water Plan Frame Work",* State Water Resource Agency, UP.

55. Siegel, S.M. 2015. *"Let there be water",* 178 Fifth Avenue, New York, St. Martin's Press.

56. Rural Development Unit, South Asian Region, World Bank and Central Ground Board, Ministry of Water Resources, Government of India. 1998. *"Ground Water Regulation and Management Report", 26.*

57. Chaurasia P. R. and Subhash. 2021. *"Bundelkhand Water Woes: Paradigm Shift is needed in the Strategy"*, Journal of Institution of Engineers (India), Series A, Published online: 02 January 2021. http://doi.org/10.1007/s40030-020-00496-8

Reviews

The book presents comprehensive information and analysis of irrigation water issues in the state of Uttar Pradesh, India. It covers surface water, groundwater, stressed areas, the river system, groundwater basins/aquifer system, etc. in a lucid manner. An overview of irrigation infrastructure development with time gives the present status. Analysis of irrigation potential created, available surface and groundwater resources with data for objective planning and effective utilization are useful. Further, analysis of water-logged areas adds value to understand the comprehensive picture.

Topics like water management initiatives in the state, present challenges, the role of forest/ green cover on water availability, and mitigating climate change look at futuristic planning to ensure water availability to crops to the extent possible. In my opinion, the book might be useful for researchers, planners, besides field functionaries and student libraries. Its availability on notionpress.com, Amazon India, USA, UK, Australia, besides Kindle book store makes it an easily accessible book.

Dated: March 31, 2021,

Bangali Baboo

ARS, M.Tech. (Ag. Engg.), Ph.D., (Mech. Engg.)
Former National Director, NAIP, ICAR,
New Delhi
Former Director, ICAR-IINRG, Ranchi
Former Head of Division, ICAR-CIPHET, Ludhiana

Water Development and Management in UP is the presentation of the current situation through the lens of two senior technocrats who have been in the position of authority for over two decades. Distinctly different from the layout of books written by academics, this has more data and opinions of insiders. This could set a trend for technocrats to bring out more such books. Young researchers should use this book to identify new areas of research that address the current problems of the State.

I wish the authors do not stop with this book but play a pro-active role in addressing bottlenecks that plague good schemes. My wishes are for an active role, post-retirement.

Dated: April 4, 2021,   **Dr. K. A. S. Mani**

Ground Water Consultant, World Bank